# Lagomorpha Characteristics

*Edited by María-José Argente, María de la Luz García Pardo and Kevin P. Dalton*

Published in London, United Kingdom

IntechOpen

*Supporting open minds since 2005*

Lagomorpha Characteristics
http://dx.doi.org/10.5772/intechopen.82908
Edited by María-José Argente, María de la Luz García Pardo and Kevin P. Dalton

Contributors
Dennis Osarenren Asemota, Arierhire Michael Orheruata, Margarida Duarte, Carina Carvalho, Fabio Abade dos Santos, Arantxa Villagrá García, Kevin P. Dalton, Ana Podadera, José Manuel Martin Alonso, Inés Calonge Sanz, Ángel Luis Álvarez Rodríguez, Rosa Casais, Francisco Parra, María-José Carrascosa Argente, María Luz Garcia Pardo, Mariam Pascual, Ernesto A. Gomez, Jéssica Monteiro, Madalena Monteiro, Paulo Melo Carvalho, Paula Mendonça, Patrícia Tavares Santos, Pedro C. Melo



First published in London, United Kingdom, 2021 by IntechOpen
IntechOpen is the global imprint of INTECHOPEN LIMITED, registered in England and Wales, registration number: 11086078, 5 Princes Gate Court, London, SW7 2QJ, United Kingdom
Printed in Croatia

British Library Cataloguing-in-Publication Data
A catalogue record for this book is available from the British Library

Additional hard and PDF copies can be obtained from orders@intechopen.com

Lagomorpha Characteristics
Edited by María-José Argente, María de la Luz García Pardo and Kevin P. Dalton
p. cm.
Print ISBN 978-1-78984-291-3
Online ISBN 978-1-78984-292-0
eBook (PDF) ISBN 978-1-78985-090-1

For EU product safety concerns:
IN TECH d.o.o., Prolaz Marije Krucifikse Kozulić 3, 51000 Rijeka, Croatia, info@intechopen.com or visit our website at intechopen.com.

# Meet the editors

María-José Argente studied Agriculture Engineering and obtained her Ph.D. with a thesis about rabbit genetics at Universitat Politècnica de València, Spain. She is an associate professor of Animal Breeding and Genetics at Miguel Hernández University of Elche, Spain. She has worked at the University of Edinburgh, University of Thessaloniki, University of Nitra, and the University of Blida. Her career has focused on the genetic improvement of litter size in rabbits using novel methods such as selecting uterine capacity and litter size environmental variability, with a recent interest in correlated response in susceptibility to stress and diseases as well as the role of gut microbiota on animal welfare. She has published more than 50 papers in scientifically indexed journals and 200 contributions to congresses. Details of her publications can be found at https://orcid.org/0000-0002-4541-3293.

María de la Luz García Pardo is an agricultural engineer from Universitat Politècnica de València, Spain. She has a Ph.D. in Animal Genetics. Currently, she is a lecturer at the Agrofood Technology Department of Miguel Hernández University, Spain. Her research is focused on genetics and reproduction in rabbits. The major goal of her research is the genetics of litter size through novel methods such as selection by the environmental sensibility of litter size, with forays into the field of animal welfare by analysing the impact on the susceptibility to diseases and stress of the does. Details of her publications can be found at https://orcid.org/0000-0001-9504-8290.

Kevin P. Dalton is a molecular virologist and assistant lecturer in the Department of Biochemistry and Molecular Biology at the Universidad de Oviedo, and member of the Instituto Universitario de Biotecnología de Asturias (IUBA) – University Institute of Biotechnology of Asturias. His research interests focus on the molecular characterization of viruses that infect rabbits. His laboratory has published many articles on both rabbit hemorrhagic disease virus and myxoma virus contributing to the discovery of emergent variants of both in recent years. The main goal of his work is to develop improved diagnostics and vaccines to help control these deadly pathogens. Details of his publications can be found at https://orcid.org/0000-0002-7086-1979

# Contents

# Preface

The order Lagomorpha includes two families, Ochotonidae with one genus (*Ochotana*, also known as pikas) and the Leporidae family with 11 genera (*Bunolagus, Brachylangus, Capralagus, Lepus, Nesolagus, Oryctolagus, Pentalagus, Poeladus, Pronolagus, Romerolagus and Sylvilagus*). Lagomorphs are widespread on all continents often due to human intervention, with the exception of Antarctica. Lagomorphs are relatively small herbivore mammals with a second pair of incisor teeth behind the first pair in the upper jaw. Coprophagia is a normal occurrence in lagomorphs, providing the animal with essential amino acids and vitamins that are synthesized by gut microbiota. There is another particular feature of the order in that ovulation is induced by copulation. Moreover, the reproductive cycle is short due to the fact that gestation lasts between 30 days, in pikas and rabbits, and 40 days in hares. In addition, females may have from two to three litters in pikas and hares and up to five litters a year in rabbits.

*Lepus* and *Oryctolagus* genera are highlighted within the Leporidae family, since all known 32 hare species come from the same genus (*Lepus*) and all European domestic rabbit breeds originated from *Oryctolagus* (*O. cuniculus* species). Both sporting and commercial use of leporids are widespread. Hares and rabbits are considered traditionally favorite species as small wild game. Additionally, there is an important commercial activity surrounding rabbit production, in order to produce meat, wool and fur, and as pets. Rabbit production is of great economic importance for the meat industry in all Mediterranean countries.

This book focuses on hares and rabbits. Chapter 1 opens with a general description of the habitat, morphological and reproductive characteristics of hares, as well as the main infectious diseases that can threaten their conservation in Europe. In relation to health, Chapter 2 addresses the molecular study of main viral diseases affecting only lagomorphs that affect wild rabbit and hare populations and produce important economic losses on domestic farms. The following chapters help the reader to understand the key items in rabbit meat production. In this regard, we would like to highlight that the rabbit meat industry is increasingly concerned about animal welfare on farms and has made marked improvements in this regard. Considering that housing, handling and health conditions directly affect animal welfare, Chapter 3 examines the optimal design for rabbit cages in terms of space, enrichment and materials, in order to provide better welfare conditions for animals. Chapter 4 examines the effects of temperature and humidity on rabbit breeding production, since the temperature-humidity index is useful to measure the risk of heat stress in animals. Chapter 5 analyzes the main traits for pecuniary interest in rabbit breeding, and the specialized sire and dam lines commercially available to the rabbit farmers. Finally, Chapter 6 develops a practical example with actual data to calculate profitability on a rabbit farm.

**María-José Argente and María de la Luz García**
Departamento de Tecnología Agroalimentaria,
Centro de Investigación e Innovación Agroalimentaria y Agroambiental,
Universidad Miguel Hernández de Elche,
Orihuela, Spain

**Kevin P. Dalton**
Departamento de Bioquímica y Biología Molecular,
Universidad de Oviedo,
Oviedo, Spain

Chapter 1

# The Health and Future of the Six Hare Species in Europe: A Closer Look at the Iberian Hare

*Margarida D. Duarte, Carina L. Carvalho, Fábio Abade dos Santos, Jéssica Monteiro, Madalena Monteiro, Paulo Melo Carvalho, Paula Mendonça, Patrícia Tavares Santos and Pedro C. Melo*

## Abstract

Although there are around 40 species of hares in the world divided into three different genera (*Lepus*, *Caprolagus*, and *Pronolagus*), only six species inhabit Europe, all belonging to genus *Lepus*. The conservation status of these six species was recently revised in the International Conservation Union (IUCN) Red List of Threatened Species. *Lepus castroviejoi* and *L. corsicanus* were attributed the status of "vulnerable". The other four species, *L. europaeus*, *L. timidus*, *L. capensis*, and *L. granatensis*, were considered of "least concern" although a declining trend was recognized for the last two species' wild populations. Here we review the major threats to the hare species in Europe, with emphasis on infectious diseases. Furthermore, we present the sanitary data regarding the Iberian hare populations from Portugal, which were severely affected by the emergence of a naturally occurring recombinant myxoma virus (MYXV), first reported in mid-2018. The recent detection in 2019 of a leporid herpesvirus (LeHV-5), which pathogenicity appears to be exacerbated in MYXV-infected hares, brings additional concerns to the health and conservation of the Iberian hare.

**Keywords:** hare species, Iberian hare, *Lepus granatensis*, viral diseases, myxomatosis, myxoma virus, MYXV, rabbit hemorrhagic disease virus, RHDV2, leporid herpesvirus, LeHV-5

## 1. Introduction

### 1.1 Geographic distribution in Europe

The Lagomorpha order (belonging to the Mammalia class) includes the Ochotonidae family, with one sole genus designated *Ochotona*, and the Leporidae family, with 11 genera, namely, *Pentalagus*, *Bunolagus*, *Nesolagus*, *Romerolagus*, *Brachylagus*, *Sylvilagus*, *Poelagus*, *Pronolagus*, *Caprolagus*, *Oryctolagus*, and *Lepus*.

Like the other hare species in the world, the six hare species found in Europe are small herbivorous mammals belonging to the order Lagomorpha, family Leporidae, and genus *Lepus*.

These hare species, however, have different geographical distributions (**Figure 1**). The Iberian hare (*Lepus granatensis*) is endemic to the Iberian Peninsula and is found in almost all of the territories in Portugal and in southwest Spain (**Figure 2**) [1]. Although genetically and morphologically distinct from the Mountain hare (*Lepus timidus*), in evolutionary terms these two species are closely related [2]. However, the Mountain hare is adapted to cold climates, being found in northern continental Europe, Scotland, Ireland, and the Swiss Alps [3], while

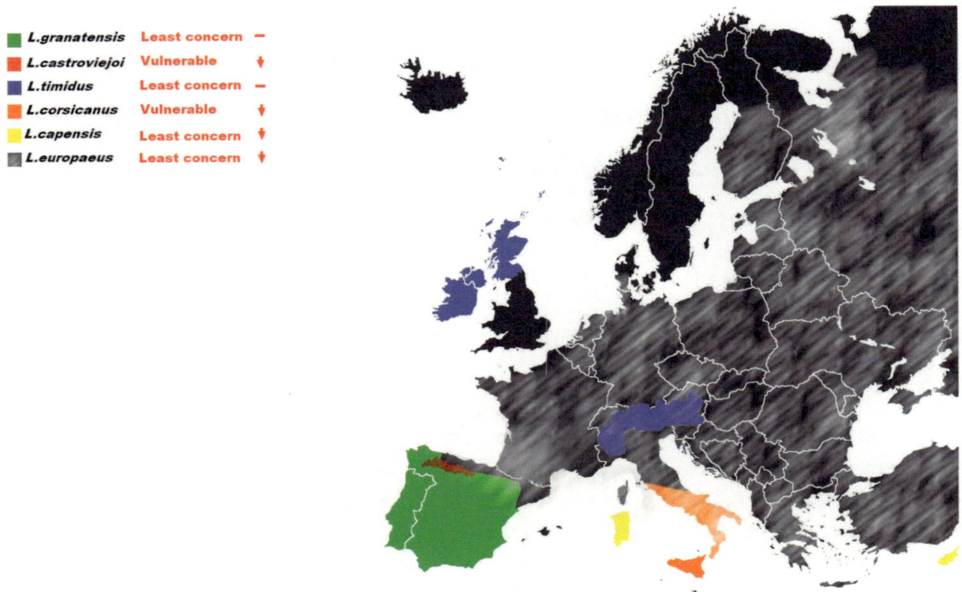


**Figure 1.**
*Geographic distribution, conservation status (International Union for Conservation of Nature, IUCN), and trends of the populations of different hare species found in Europe. The distribution was inferred from the distribution on the IUCN maps, with low precision of the geographic distribution limits. Only distribution in Europe is shown. The distributions resulting from human introductions are not represented. For more information, see the individual page for each species on the website of the International Union for Conservation of Nature (IUCN).*

**Figure 2.**
*Adult male Iberian hare (photograph by Sebastião Miguel, 2019).*

the Iberian hare's preferred habitat is composed of undergrowth plains, alternating with small areas of bush or grove for refuge [4]. In Spain, two other species are also present, namely, the Broom hare (*Lepus castroviejoi*), limited to the Castroviejo region of northern Spain [5], and the European or Brown hare (*Lepus europaeus*), the species with the widest geographical distribution, also found in the countries of Central and Eastern Europe [6]. The other two species that inhabit Europe are the Italian or Corsican hare (*Lepus corsicanus*), native to the southern coast of Italy and Sicily [7], and the Cape, Arabian, or desert hare (*Lepus capensis*), found in Sardinia. The latter originates from Africa, Asia, and the Middle East and, unlike the other five mentioned species, was introduced into Europe [8]. The six hare species and respective subspecies that inhabit Europe are identified in **Table 1**.

| Species | Subspecies | | Common names |
|---|---|---|---|
| ***Lepus europaeus*** | *L. e. europaeus* | *L. e. caspicus* | Brown hare |
| | *L. e. connori* | *L. e. creticus* | European hare |
| | *L. e. cyprius* | *L. e. cyrensis* | |
| | *L. e. hybridus* | *L. e. judeae* | |
| | *L. e. karpathorum* | *L. e. medius* | |
| | *L. e. occidentalis* | *L. e. parnassius* | |
| | *L. e. ponticus* | *L. e. rhodius* | |
| | *L. e. syriacus* | *L. e. transsylvanicus* | |
| ***Lepus timidus*** | *L. t. ainu* | *L. t. begitschevi* | Mountain hare |
| | *L. t. gichiganus* | *L. t. hibernicus* | Blue hare |
| | *L. t. kamtschaticus* | *L. t. kolymensis* | Tundra hare |
| | *L. t. kozhevnikovi* | *L. t. lugubris* | Variable hare |
| | *L. t. mordeni* | *L. t. orii* | White hare |
| | *L. t. scoticus* | *L. t. sibiricorum* | Snow hare |
| | *L.t. sylvaticus* | *L. t. timidus* | Alpine hare |
| | *L. t. transbaicalicus* | *L. t. varronis* | Irish hare |
| ***Lepus granatensis*** | *L. g. granatensis* | | Granada hare |
| | *L. g. gallaecius* | | Liebre ibérica (Spanish) |
| | *L. g. solisi* | | Lebre-ibérica (Portuguese) |
| ***L.castroviejoi*** | No subspecies | | Broom hare |
| ***Lepus capensis*** | **South Africa group** | | Sardinian hare |
| | *L. c. capensis* | *L. c. aquilo* | |
| | *L. c. carpi* | *L. c. granti* | |
| | **East Africa group** | | |
| | *L. c. aegyptius* | *L. c. hawkeri* | |
| | *L. c. isabellinus* | *L. c. sinaiticus* | |
| | **East Africa group** | | |
| | *L. c. arabicus* | | |
| | **North West Africa group** | | |
| | *L. c. atlanticus* | *L. c. mediterraneus* | |
| | *L. c. schlumbergeri* | *L. c. whitakeri* | |
| ***Lepus corsicanus*** | No subspecies. Recently considered to be a distinct species from *L. europaeus* | | Corsican hare |
| | | | Apennine hare |
| | | | Italian hare |

**Table 1.**
*Scientific and common names of the six hare species found in Europe. The recognizable subspecies for* Lepus europaeus *(16 subspecies),* Lepus timidus *(16 subspecies),* Lepus granatensis *(3 subspecies) and* Lepus capensis *(13 subspecies) are shown.*

Southern Europe provides, therefore, suitable habitats for the largest number of hare species. Morphologically, the six species of hare that inhabit Europe are distinguishable and follow the Bergmann rule, which establishes a direct relationship between the adults of medium size and the colder environments, for a given taxonomic group with wide geographical distribution [9].

### 1.2 Favorite habitats

The Iberian hare occupies a wide variety of habitats [2], namely, coastal dunes, wet mountain forests, and dry areas [10]. Like the other species found in Europe, such as the European hare, generally it does not need open water to sustain its metabolism [11]. Besides the open fields, the greater species densities are registered in intensive agricultural areas [12, 13] such as olive tree, sunflower fields, and vineyards [14].

### 1.3 Morphological characteristics

The Iberian hare is smaller than the other sympatric species, namely, the European hare and Broom hare, with mean body weight ranging from 2.0 to 2.6 kg [15]. Females are bigger than the males [10]. The Iberian hare has an extensive white ventral area that extends partially to the forefeet and hindfeet. This species has an evident contrast between the fur color of the back (ochraceous brown/gray-brown) and the belly (white). It has large brown eyes and long ears (with dark extremities) as a heat dissipation mechanism. The tail is also black on the dorsal surface and white on the ventral side (**Figure 3**). The hind limbs are longer than the front ones [9]. These characteristics added to a cleft lip and second pair of incisors in the upper jaw allow for the differentiation of leporids from rodents.

**Figure 3.**
*Lateral-caudal view of a juvenile male Iberian hare (photograph by Margarida Duarte, 2019).*

### 1.4 Natural behavior

Hares are solitary, as they do not have a social organization nor inhabit burrows [14]. However, they can gather in groups following complex age-dependent patterns, mostly during feeding time, hence reducing predation risk and increasing feeding efficiency [16], or at the time of mating [17]. They do not have a territorial behavior, unlike other lagomorphs such as the wild rabbit [18].

Hares are active primarily during twilight and at night, though in summer they may be observed during the day [19]. During daylight, they seek refuge at the surface, in depressions a few inches deep, dug into the ground or in foxes and marmot's burrows [11].

The Iberian hare is highly specialized in camouflage and when chased by predators is capable of rapid escape, reaching around 70 km/hour [20]. It has a relatively lighter skeleton and larger heart than rabbits, which is only found in the fast-running species [9].

### 1.5 Reproduction

Reproductive parameters and breeding activities depend on the hare species and environmental conditions. A study on the reproductive strategies of genus *Lepus* compared the breeding season and litter size for distinct hare species, showing differences depending on the climatic conditions of the breeding areas. The species that occur in zones of greater latitude usually produce only a litter per year of about 6–7 young, while species in temperate climate zones have a longer reproductive period, with 3–4 litters of 2–5 leverets each. In the regions closer to the equator, there is no interruption in the reproductive period, with an average of 8-litter per year, each with 1–2 young [21].

The European brown hare, best studied due to its extensive geographic distribution, is a polyestrous seasonal breeder [22]. During the breeding peak, in the spring, mating leads to agglomerations of solitary hares, the so-called "March madness". This species produces an average of 3–5 litter per year [23]. The mean litter size appears to be dependent on the region occupied, ranging between 2.0 and 2.7, with a maximum of 6 [24]. European hares newborns have an average weight of around 100g and are fully furred, born with eyes open, and able to walk. Weaning occurs around 4–5 weeks when juvenile weight reaches around 1 kg. The European hare is fertile at around 4–5 months of age reaching maximum weight at 8 months. When in continental climates, the reproduction in the year of birth is frequent [11]. This species has an age expectancy of 8–12 years [25].

The principal breeding season of Mountain hare occurs from February to September. The gestation period is about 42 days, but this inter-birth interval can be of 36 days in case of superfetation [11]. The species has a mean litter size between 1.9 and 2.1 with a maximum of 5 leverets.

The reproduction period of Iberian hare occurs throughout the year, although there is a certain seasonality in its reproductive activity, peaking in March and April with a minimum in autumn [24]. The onset of sexual activity is not season dependent but rather depends on the size of the animals [10]. The Iberian hare reproductive strategy, of continuous procreation [24], is concordant with smaller litters and longer breeding seasons [26]. Gestation period is also around 42 days. The seasonal trend in the population of young depends on the percentage of pregnant females and litter size. The most frequent gestations involve one or two fetuses; however, litter size may range from one to seven leverets, the largest litter size reported in the wild [11, 27]. The mean annual litter size was estimated in 2.1 leverets per litter [27]. Based on embryo

**Figure 4.**
*Leveret of Iberian hare approximately one month of age (photograph by Margarida Duarte, 2019).*

counts, the mean litter size was 1.58 (range 1–4). Annual changes in the environment impact on the reproduction of the Iberian hare causing seasonal variations [27]. Newborn Iberian hares are also fully furred, born with eyes open, and able to walk [11] (**Figure 4**). These characteristics differentiate hares from other lagomorphs [28]. The mean weight of newborn leverets in captivity is 128.6 g (range 123–140 g) [11].

### 1.6 Ecological relevance

Hares have an ecological substantial importance as prey of several species like the golden eagle (*Aquila chrysaetos*), the European wildcat (*Felis silvestris*), the red fox (*Vulpes vulpes*), the Eurasian eagle owl (*Bubo bubo*), among others [29]. Due to the decrease in the number of wild rabbits (*Oryctolagus cuniculus*), the Iberian hare also plays an important role as prey for predators such as the *vulnerable* imperial eagle (*Aquila adalberti*) [30–32].

### 1.7 Importance as a small game species

As a game species, the Iberian hare is much appreciated by Portuguese and Spanish hunters (**Figure 5**). In Portugal, hare hunting is permitted from September to February by different modalities, namely, "salto" and "batida" (the name of two hunting processes in the Portuguese territory), and also standing, coursing, and falconry (Article 93, Decree-Law No. 202/2004 of 18 August); the last two are only allowed between January and February. The ability of high-speed endurance running is used in hare coursing (**Figure 6**), a modality that has led to the selection of the greyhound breed.

Although there is no hare population census-supported data from Portugal, the trends from the National Gamebag Census indicate a reduction in the Iberian hare populations in the recent decades that has accompanied the decline in the wild rabbit populations [33]. This decrease resulted from the combined and cumulative effect of several environmental factors that simultaneously affected the wild rabbit and the Iberian hare, along with the emergence of infectious diseases, namely, myxomatosis and rabbit hemorrhagic disease (RHD).

**Figure 5.**
*Portuguese hunter (Jacinto Amaro, President of the Portuguese Hunting Federation, Fencaça), collecting an Iberian hare from his dog (German Shorthaired Pointer x English Setter), Avis, Alentejo, 21st December 2017 (photograph by Margarida Duarte).*

(A)

(B)

**Figure 6.**
*Top: João Grave and his team performing field recognition before a race (Herdade da Bala, Évora, Alentejo, 20th February, 2016). Bottom: Greyhound dogs chasing an hare during a coursing race (Alentejo, 20th February, 2016). Photographs by José d' Oliveira e Sousa.*

## 2. The top threats to hares in Europe

Most hare species have been subjected to a multitude of threats, which consequently have led to the reduction of the wild populations.

The International Union for Conservation of Nature (IUCN) has recognized as threatening factors to hares in Europe, the loss of habitats due to changes in the agriculture practices and development of urban areas, overhunting, poaching, trapping, agriculture pollution, human intrusions, and disturbances due to recreation activities. The impact of invasive non-native or problematic native species with consequent competition and hybridism has also been pointed out as treats to hare species.

The fragility of the hare species is also related to its own idiosyncrasies, such as the biology of its reproduction. The relatively small size of the litter and the characteristics of the shelters constructed above ground expose the juveniles to much higher predation rates than do rabbits.

Juvenile mortality is considered the most critical factor in the population dynamics. The hares' abundance is directly related to female breeding success and to juvenile survival rates [34], both directly dependent on habitat suitability [14]. The Iberian hare population dynamics is also greatly affected by food availability [35]. The highest juvenile mortality is observed after the maximum reproductive intensity period [14]. This mortality is due to, among other factors, agricultural landing, diseases, and predation [11]. Juvenile mortality of the European hare may reach 90% [36]. Data on the Iberian hare suggest that nearly 60% of the young die, corresponding to an increase of 40% in the population numbers [14]. More recently, prenatal mortality was estimated between 18% and 21% [24, 26]. One study refers to a minimum annual survival rate in young of 27.91% [27].

Below we detail some of the main threats to hares in Europe.

### 2.1 Habitat loss

Many factors have cumulatively led to ecosystem modifications and to the deterioration of the hares' preferential habitats. The changes in agricultural practices, namely, by the cultivation of annual and perennial non-timber crops, have played a major role in habitat reduction. In addition, the expansion of urban and industrial areas as well as of roads and railroads brought limitations and barriers to the natural habitats and movements of wild species [1, 3, 5–8].

A meta-study involving 12 European countries concluded that the primary cause of the European hare decline since 1910 was agricultural intensification [11]. The average density of the European hare in the original habitats was two individual per 100 ha, although densities of 275/100 ha have already been recorded in territories with favourable conditions [11]. The territory range is directly dependent of the agricultural intensity and can be less of 20 ha (well-structured habitat) or up to 300 ha in habitats subject to intensive agriculture [11].

In the case of the Iberian hare, climate changes and reforestation of old cultures with the densification of open scrubland areas have been contributing to lower habitat suitability [1].

### 2.2 Diseases

#### *2.2.1 Management of wildlife diseases*

Management strategies for wildlife disease encompass the prevention of introduction and spread of new diseases and control or eradication of an existing disease.

However, management of hares' diseases is hampered for several reasons. The high resistance of some pathogens in the environment, where they may persist infectious for long periods, the lack of identification of putative species' reservoirs, and their dissemination via arthropod vectors, such as the European brown hare syndrome virus (EBHSV), RHDV, RHDV2, and MYXV, potentiate new infections and make control very difficult.

Viral dissemination of leporid diseases has often been linked to anthropogenic activities [37, 38]. Changes in human activities, such as the introduction of cleaning and disinfection practices after hunting, the proper disposal of animal by-products, the periodic deworming of dogs, and the restriction of hare translocations to prevent disease spreading, can be easily implemented and impact positively on the recovery of the populations.

Immunization of wild populations, through oral bait vaccines, has been successful in the control of a few diseases such as rabies [39] and Aujeszky disease [40]. However, immunization is best suited for microparasitic exogenous infections with a low reproductive rate and in populations which have a low turnover [41], which is not the case of wild leporids. Several attempts to produce vaccines against RHDV for wild rabbits have been made in the past [42, 43].

Disease surveillance programs of wild animal populations are particularly crucial to obtain data regarding the animal health. The frequent movements of people through territories previously not occupied by man and the recurrent translocation of animals for hunting and conservation purposes increase the contact between wild and domestic animals and humans.

#### *2.2.2 Importance of a conclusive diagnosis in decision-making*

Despite its limitations, the implementation of appropriate infection control measures must always be supported by a conclusive diagnosis, for which laboratory testing is essential.

The clinical evaluation of sick animals for diagnosis and research purposes is particularly difficult in wild species, where samples usually reach the laboratories as cadavers or organ samples. In these cases, histopathology provides the only possible bridge to understand the physiopathology of the disease. However, many factors compromise the rigor of the histopathological evaluation, such as an advanced degree of autolysis, which is common in wild species, and the consequences of sample freezing, often done due to transportation and logistic difficulties. Nonetheless, an exhaustive and systematic necropsy is the basis for a successful and complete laboratory diagnosis.

Isolation of pathogens, namely, of viruses by multiplication in sensitive cell lines, facilitates diagnosis and research. However, some pathogens of leporids are not cultivable *in vitro*, such as RHDV [44] and RHDV2 [45].

Molecular methods, in use for many decades, provide the possibility of testing many samples in a few hours with high specificity and sensibility.

Serologic examination of wild species allows the assessment of the previous contact of a population with a particular pathogen (herd immunity).

#### *2.2.3 Relevant hare pathogens*

Until recently, unlike rabbits, infectious diseases were not considered a major threat to hares, except for the European brown hare syndrome (EBHS) [46].

Pathogens that infect hares may have an impact at individual or population level. Some are also zoonotic causing disease in humans.

Staphylococcosis (caused by the bacterium *Staphylococcus aureus*) and toxoplasmosis (caused by the protozoan *Toxoplasma gondii*) may affect hares but generally do not have a major impact on their health from a population perspective. Cestodal or tapeworm infections within this group comprise *Paranoplocephala wimerosa*, *Andrya cuniculi*, *Andrya rhopalocephala*, *Cittotaenia denticulata*, *Mosgovoyia pectinata*, and *Mosgovoyia ctenoides*, all causing catarrhal enteritis with malabsorption in severe cases [47].

Examples of diseases that may have an impact at the population level are pasteurellosis, outbreaks of which, despite residing within the upper respiratory tract of most animals, can kill up to 80% of the population. Hares are also particularly vulnerable to coccidial infections. It is believed that Coccidia play an important role in leveret's mortality rate but it also affects adults. Several species of *Eimeria* were reported in hares, namely, *E. europea, E. hungarica, E. robertsoni, E. robertsoni, E. septentrionalis, E. stefanskii* and *E. townsendii*, which can cause severe catarrhal enteritis and gaseous distension of the gut being found within the epithelial lining of the intestines. *E. stiedae* is limited to the liver and is less important in hares than in rabbits [47].

The gastrointestinal nematode, *Graphidium strigosum*, resides within the stomach of up to 40–60% of hares and in massive infections may cause anemia. In addition, *Trichostrongylus retortaeformis*, a nematode that causes catarrhal enteritis, and *Trichuris leporis*, found inside the cecum which produces toxic metabolites responsible for necrotic lesions within the gut wall, may also be present in high percentages in hare populations, namely, 75.8 and 39.8%, respectively [47, 48]. As for lungworm parasitosis, there are reports of 42–60% of *Protostrongylus commutatus* infection within a hare population [47]. In severe cases, animals present dyspnea and seromucosal nasal discharge due to catarrhal pneumonia and pleuritis. Lungworm infections seem to predispose to contact with bacterial disease [47–49]. Kornaś et al. [50] found a higher prevalence of nematode infection among adult hares than in juveniles from Southern Poland.

The typical examples of the zoonotic disease group are tularemia, a bacterial disease caused by *Francisella tularensis* [51], and leishmaniasis, caused by protozoan parasites of the genus *Leishmania* that are transmitted through the bite of female sand flies [52]. *Francisella tularensis* is one of the most virulent microorganisms presently known as few as 10 microorganisms can cause potentially fatal disease in man and animals (reviewed in [51]). The most important pathogenic subspecies are *F. tularensis* subsp. *tularensis* (Type A) that occurs usually in North America and *F. tularensis* subsp. *holarctica* (Type B) that occurs throughout the northern hemisphere (reviewed in [51]) and has been described in Iberia both in wild leporids [53] and humans [54]. Wild lagomorphs are one of the main reservoirs of *F. tularensis* in nature and are considered suitable sentinels for disease surveillance (reviewed in [51]). *Leishmania infantum* is responsible for both visceral and cutaneous zoonotic leishmaniasis in the Mediterranean Basin. Iberian hares were associated with an outbreak of 260 human cases of leishmaniasis affecting metropolitan Madrid, Spain, suggesting that hares may have an unexpected role in the epidemiology of *L. infantum* in Spain [52].

Brucellosis is another important zoonosis that can infect hares. In this species, it is caused by *Brucella suis* biovar 2 which can infect other wild or domestic animals and humans [47, 55]. Additionally, hares can also be infected with the zoonotic important *Echinococcus granulosus* and *Echinococcus multilocularis* [56–58]. Pseudotuberculosis is another typical disease of lagomorphs with zoonotic potential. It is caused by *Yersinia pseudotuberculosis* and is one of the most important lethal infections in hare with population losses of up to 50% [47, 59, 60]. Generally, the zoonotic pathogens do not have a major impact on hares, which act mainly as reservoirs for humans [51, 52, 55].

Some pathogens, such as *Taenia pisiformis* (a cestode parasitosis) [61], and a few viruses, namely, EBHSV [17], rabbit hemorrhagic disease type 2 (RHDV2) [62–67], and the new natural recombinant myxoma virus [68–70], cause potentially devastating diseases in hares, constituting real threats to the preservation of the wild populations. These pathogens are described in more detail below.

*2.2.3.1 Cysticercosis*

*Taenia pisiformis* is known to cause a typical parasitosis of lagomorphs known as cysticercosis [33, 71]. The larval stage of this parasitic cestode is found particularly in rabbits and hares, having been described in the European brown hare [61] and in the Iberian hare [72]. Generally, cysticercosis does not give rise to clinically relevant signs in lagomorphs [61]. Light infections are unapparent, although heavy infections can cause abdominal distension and discomfort [73, 74]. Notwithstanding, a negative relation between cysticercosis and kidney fat index in Iberian hare and loss of prolificacy in New Zealand rabbits has been described [61, 75, 76].

There is little information on the prevalence and species diversity of cestode infections in rabbits because of their low pathogenicity and the limited opportunities available to diagnose infection [74]. In the European brown hare, the prevalence of *T. pisiformis* found in northern Italy was 14.8% (8/54) in 2013 and 3.28% (2/61) in 2015 [61]. In Portugal, cysticercosis has been frequently observed in the Iberian hare in some geographic areas from the south.

A combination of pathological-, parasitological-, and molecular-based techniques is usually employed for diagnostic purposes [74].

The presence of mature cysticerci within the abdominal cavity, the most common clinical presentation, can be observed during postmortem examination [77–79]. Yellowish-white parasitic cysts (2–18 mm) are observed in diverse locations, namely, in the peritoneum and liver surface. In massive infections, cysts may be found in the thorax, affecting the mediastinal space and pulmonary parenchyma, scrotum, small and large intestine, and renal capsule (**Figure 7**). Moderate to severe hepatomegaly may be observed.

Vesicles are covered by a membrane (thickness about 1 mm) containing clear liquid and an invaginated scolex of white cysticerci [80, 81]. In some cases, the migratory path of the larvae in the liver can be visualized as lighter colored areas. A host reaction to the parasite, with moderate interstitial lymphocytic hepatitis is observed. Liver parenchyma around protoscolex sections of cysts is usually surrounded by a

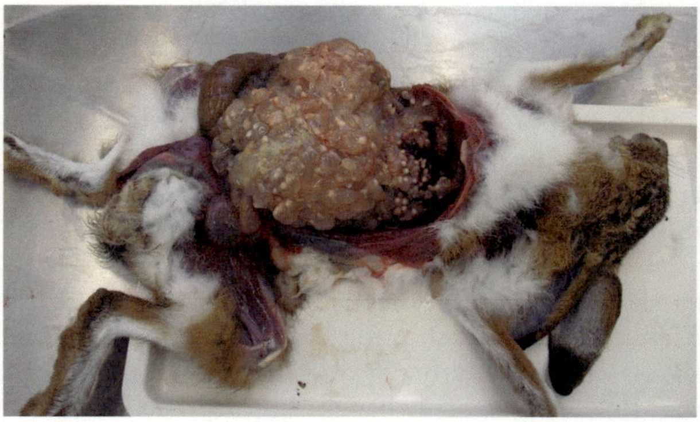

**Figure 7.**
*Macroscopic view of the abdomen and thorax cavities of an adult female hare with numerous cysticerci (the larval stage of Taenia pisiformis), attached to the serosa of the intestines and to the liver surface (photograph by the INIAV I.P. Pathology team).*

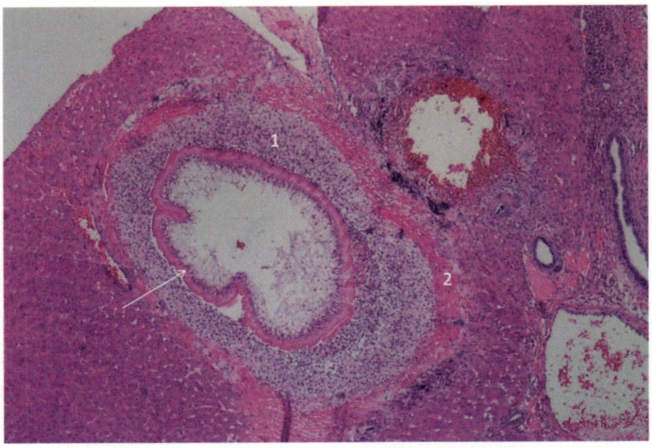

**Figure 8.**
*Microscopic view of a cysticercus vesicle (arrow), in the liver of an Iberian hare. Inflammatory cell infiltration (1) and peripheral necrosis (2) around the vesicle. HE, 40× (photograph by the INIAV I.P. Pathology team).*

fibrous reaction or granulomatous inflammation with multinucleated giant cells, macrophages, plasma cells, and many eosinophils (**Figure 8**). Multifocal to disseminated interstitial chronic hepatitis with diffuse biliary stasis may be observed [74].

At the parasitological examination, cysticerci appear as small, transparent, fluid-filled cysts with a broad anterior and narrow, tail-like posterior. Examination by light microscopy using staining techniques can reveal features consistent with described morphologies of *T. pisiformis* cysticerci, namely, corrugated tegument, apical tegument invagination, and invagination canal [74, 82–85].

The molecular diagnosis of *T. pisiformis* cysticercosis can be performed using a generic pair of primers described by Boubaker et al. [86]. This pair of primers was designed to amplify the *Taeniidae* mitochondrial 16S ribosomal RNA gene and is suitable for amplifying and distinguishing through sequence analysis 13 separate cestode species and theoretically for distinguishing further 10 cestode species, predominantly from the family *Taeniidae* [86].

#### *2.2.3.2 European brown hare syndrome*

European brown hare syndrome (EBHS) is caused by a single-stranded RNA non-enveloped virus (EBHSV) belonging to the *Lagovirus* genus, family Caliciviridae, which induces a disease in the European brown hare similar to rabbit hemorrhagic disease (RHD) in rabbits, characterized by hemorrhages in several organs and liver necrosis. Despite EBHSV emergence was recognised in the 1980s in the north of Europe [46], Duff et al. [87] reported descriptions of lesions consistent with EBHS in hares since 1976 in England. In addition, Lenghaus et al. [88] revealed that hunters in Scandinavia knew of the disease in the early 1970s. Until now, EBHSV is restricted to Europe having been registered in the European hare in Sweden [46], Italy [89], the United Kingdom [87, 90], France [91], Poland [92], Greece [93], and Slovakia [94].

EBHS causes severe necrotic hepatitis in both wild and captive hares (European brown hares and mountain hares) [46, 93, 95]. The infection has significant similarities to RHD in its epidemiology, symptomatology, and pathology [93, 96], being characterized by rapid progression, mild nervous symptoms (including depression, muscular tremors, and incoordination), presence of sero-haemorrhagic liquid at the nostrils, congestion and extensive haemorrhages on the lungs and on serosa and mucosa, severe necrotic hepatitis, and congestion of the spleen and kidneys [90, 92, 93, 96].

Severe congestion of the trachea may also be present. Splenomegaly and dark red-black discoloration of the spleen, kidney congestion, and hepatomegaly may also be present. Liver is usually friable and discolored [93, 97]. Death occurs within 3 days after the onset of clinical signs, and mortality rates are extremely high and can reach 100% [90, 92].

Microscopically, necrosis may affect the whole hepatic lobule or be confined to peripheral or periportal areas. Hepatocytes nucleus appear swollen and lytic or may have completely disappeared. Coagulation necrosis can occur. Periportal and midzonal hepatocellular necrosis is the most consistent histopathologic finding. Nuclear degradation appears to be rare in EBHS contrary to what happens in RHD. The hepatocytes frequently contain fine iron granules.

Other findings are marked hyperemia irregularly distributed over the hepatic lobule and disruption of sinusoids with hemorrhage. In the necrotic and hemorrhagic areas, neutrophil infiltration and mesenchymal cell proliferation are observed.

Moderate dilatation of segments of the proximal and distal tubules with flattened epithelium and focal hydropic degeneration of the proximal tubules may be present. The tubules can contain pale eosinophilic protein casts of a slightly granular appearance. Congestion of spleen red pulp has been described. White follicles can appear depleted of lymphocytes, with karyorrhexis or pyknosis of B and T cells. Follicular hyperplasia can also occur [93, 97].

As with other lagoviruses, EBHSV cannot be isolated or propagated in rabbit and hare primary cell cultures or in cell lines (RK-13, PK-15, FL) [46, 98, 99]. The virus can be detected usually in the liver, which contains high viral loads, by electron microscopy, which shows 30–35 nm virions indistinguishable from those of RHDV [90, 100]. Other methods for virus detection include the hemagglutination (HA) test, using human type "A" or "O" red blood cells [46, 100], enzyme-linked immunosorbent assay (ELISA)-based methods [46, 90, 96, 98], and RT-PCR. Bascuñana et al. [99] developed two conventional RT-PCR assays for the detection and differentiation of RHDV and EBHSV, both able to detect as few as 10 copies of cloned viral genomic fragments, with no cross-amplification between the two viruses. The system can be used for amplification of VP60 genomic sequences from fresh and fixed tissues. Primers were selected from similar regions of the VP60 genes to amplify a fragment of 316 nucleotides from the genome of RHDV and a region of 265 nucleotides from the genome of EBHSV. In addition, Le Gall-Reculé et al. [101] developed an immunocapture (IC)-RT-PCR for EBHSV diagnosis that can be carried out directly with the liver exudate. Viral particles present in the sample are captured by specific antibodies immobilized on a microtitration plate. After enzymatic disruption of virus-antibody complexes, viral RNA is released and subjected to RT-PCR. The assay combines the rapidity of an ELISA test (because immunocapture and the RT reaction are carried out in the same microtiter plate) and the sensitivity of PCR, being suitable for the processing of large numbers of samples and phylogenetic studies. In 2011, Zexiao et al. [102] also developed a RT-PCR for the detection of EBHSV with good specificity and sensitivity, able to detect about 50 copies of cloned viral genomic fragments (pGM-T-EBHSV), with no amplifications for RHDV.

#### *2.2.3.3 Rabbit hemorrhagic disease*

RHD is caused by two types of lagoviruses, namely, RHDV that emerged in China in 1986 in domestic rabbits [103] and RHDV2 (RHDVb or GI.2), a virus genetically related to but distinct from RHDV, that emerged in France in 2010 in rabbits [104]. RHDV2 quickly replaced the circulating strains of RHDV in most European countries, both in the wild and domestic populations. In the Iberian hare, Lopes et al. [105] reported RHDV only in two animals during a retrospective study. However, in the last decade, several cases of RHDV2 disease have been reported in the European hare in France (2013) [64], Spain (2014), and Italy (2012) [65], in the United Kingdom (2018 and 2019) [67], in Australia (2015) [106], in Sweden (2016 and 2017) [38], and in the Netherlands (2017) (https://www.dwhc.nl/en/haas-rhdv-2-nederland/). RHDV2 was also reported in the Cape hare on the island of Sardinia (2011) [63] and in the Italian hare, in Sicily (2012) [34]. More recently (in 2019), cases of RHDV2 infection have been reported in the Mountain hare in Scotland [107] and Ireland (https://www.oie.int/wahis_2/public/wahid.php/Reviewreport/Review?page_refer=MapFullEventReport&reportid=31386).

However, RHDV2 infections have never been described in the Iberian hare.

The clinical evolution of RHD can be peracute, acute, subacute or chronic. The clinical manifestations are mainly present in the acute form of infection though in the subacute form, similar but milder signs can be observed. In the peracute form, usually there are no clinical signs of disease [108]. Chronic and subacute forms are more frequent in RHDV2 infections [109], which also differ from RHDV in affecting rabbit kits as young as 11 days old that develop disease [104, 109, 110].

The incubation period may vary between 1 and 5 days. Death may occur 12–36 hours after the onset of fever (>40°C). The clinical signs observed include prostration, apathy, convulsion, ataxia, paralysis, opisthotonos, paddling, anorexia, dullness, groans and cries, dyspnea, frothy and bloody nasal discharge, and cyanosis of mucous membranes. Subacute cases may present with malaise, mild anorexia, apathy, weight loss, and jaundice [108].

RHDV2-infected hares exhibited clinical signs and lesions similar to those induced by EBHSV [64–66]. Epistaxis and a RHD-like disease was reported in the Italian hare [34] and in the Sardinian Cape hare [63].

The macroscopic findings in hares affected by RHDV2 include extensor rigidity, epistaxis, and hyperemia of the tracheal mucosa, where a foamy bloody mucous can be found. The liver can be pale or congested (light brown to orange-pink) or presenting a reticular enhancement pattern suggestive of zonal vacuolar hepatocellular degeneration and necrosis. Moderated to severe congestion or petechial hemorrhages (or multifocal to coalescing hemorrhage) can also be observed in the lungs and kidneys.

Microscopic lesions may comprise generalized hepatic necrosis, including coagulative necrosis and multifocal areas of lytic necrosis. Acidophilic bodies may be observed in the hepatocytes. Other findings may include fatty degeneration of hepatocytes and periportal mononuclear infiltration. Degeneration and necrosis of kidney proximal tubules cells or acute tubular nephrosis may also be registered. In some cases, it is possible to observe moderated lymphocytolysis in the spleen white pulp, and moderate to severe fibrin deposition and necrosis of the red pulp can be observed [62–64, 66, 67].

Different molecular assays for the detection of RHDV have been described since the late 1990s, including conventional RT-PCR assays [99, 111–113], immunocapture RT-PCR [110], real-time multiplex RT-PCR [114], and more recently, loop-mediated isothermal amplification [115] and SYBR green-based real-time PCR [116]. None is designed to specifically detect RHDV2 strains.

For the amplification of RHDV2, Le Gall-Reculé et al. [109] described specific primers which amplify a 794-bp sequence located in the C-terminal of the gene encoding the VP60 of RHDV2.

Duarte et al. [117] developed a specific real-time TaqMan RT-PCR for the detection of RHDV2. The system was designed to amplify a 127-nucleotide-long RNA region located within the *vp*60 gene, being able to detect as few as 9 RNA molecules. The method has proven a valuable tool to diagnose most of RHDV2 circulating strains and useful to monitor viral loads [118], disease progression, and vaccination efficacy [119]. The system figures in the OIE Manual as a recommended method for nucleic acid detection (https://www.oie.int/fileadmin/Home/eng/Health_standards/tahm/3.06.02_RHD.pdf). Nearly 5 years since it was developed, the method remains suitable for the detection of the circulating field strains [120].

More recently, Dalton et al. [121] developed a diagnostic real-time RT-PCR for the detection of RHDVb strains targeting a 91-bp amplicon within the VP60 protein that covers nucleotides (nts) 6190–6280. The RT-PCR is carried out as a duplex using the endogenous amplification control of the beta-actin gene from a commercial kit (EXOPOL S.L.). The same authors also designed a conventional RT-PCR for the differentiation of RHDV2 strains from RHDV2 recombinants by subsequent sequencing of the amplicon. Primers were designed to cover a 449-nt region from the 3′ region of the RHDV polymerase (3D) to the sg promoter region. Degenerate primers were designed to bind at positions 4837–4856 and 5266–5286 in the RHDV2 genome [121].

#### *2.2.3.4 Myxomatosis*

Myxomatosis, caused by a double-stranded DNA enveloped virus belonging to the family *Poxviridae*, genus *Poxvirus*, was for a long time considered a disease of rabbits. Only a few sporadic cases were reported in the European brown hare in Spain in 1953 [122], France in 1954 [123], and the United Kingdom in 2014 [124]. However, this scenario changed radically in 2018, when outbreaks of myxomatosis in Iberian hare occurred in Spain [125] and in Portugal [126] https://www. oie.int/wahis_2/public/wahid.php/Reviewreport/Review?page_refer=Map FullEventReport&reportid=28628, causing an alarming wave of mortality in wild populations [68, 126].

The hare myxoma virus (MYXV) is a recombinant with additional genetic material in comparison with the MYXV from rabbits. The inclusion of this supplementary material led to the disruption of one gene and to the duplication of four others. While the origin of this recombinant is not yet completely understood, some authors have suggested that the extra genes originated from a capri- or cervipoxvirus [69, 70], and others propose the translocation of genes from the rabbit myxoma genome itself [70].

In rabbits, when the typical signs and skin lesions develop, the clinical diagnosis is possible. These signs may include blepharitis, blepharoconjunctivitis, anorexia, listlessness, fever, and depression. In the dermatrophic form of myxomatosis, disseminated cutaneous and mucocutaneous lesions, accompanied by oedema, are usually present in the eyelids, nose, lips, ears, and genitals (vulva, or scrotum). Emaciation and dyspnea followed by death may occur within a few weeks [127, 128]. However, in peracute forms, sudden death may occur with no clinical signs of disease or unspecific signals such as lethargy, loss of appetite, fever, and eventually swelling of the eyelids [129]. The clinical diagnosis of the atypical amyxomatous forms of myxomatosis, characterized by minor cutaneous signs and intense respiratory distress, may be complicated as well [127, 129–131].

In hares, as for rabbits, the clinical diagnosis relies on the observation of the typical skin lesions, namely, small nodules and oedema, usually present in the eyelids, nose, and genitalia. The typical myxomas observed in the ears, eyelid, or other areas of the skin in rabbits are rarely observed in hares (**Figure 9**). Blepharoconjunctivitis and mucopurulent eye discharge are a common feature [68, 69, 126].

MYXV-infected hares can show good body condition, contrary to what is common in rabbits with myxomatosis.

Histopathological findings include epidermal hyperplasia and balonization of the epidermal keratocytes. Eosinophilic cytoplasmic inclusion bodies may be observed in the keratinocytes. In the dermis, an abundant basophilic myxoid matrix

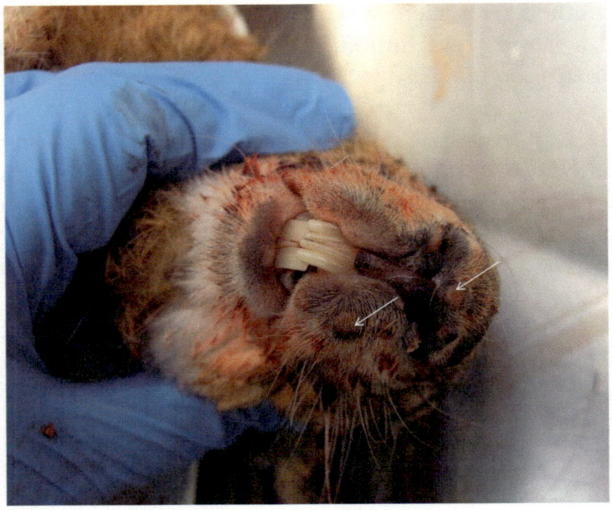

**Figure 9.**
*Iberian hare positive to myxoma virus, with nodular formations in the lips (arrows) (photograph by the INIAV I.P. Pathology team).*

admixed with oedematous areas with inflammatory infiltrates of macrophages, lymphocytes, and polymorphonuclear cells may also be seen (**Figure 10**). A severe depletion of lymphocytes may be noted in the spleen [68, 132, 133]. As for rabbits, the clinical diagnosis of the atypical form of the disease in hares may be limited.

Molecular methods provide the possibility to detect a reduced number of viral DNA copies in multiple tissue samples such as nasal and conjunctival swabs, skin oedema, myxomas, crusts, lungs, and semen. The highly conserved regions of the MYXV genome are used for primer design for PCR-based assays, allowing for detection of the circulating virus strains. Several conventional [134] and real-time PCR methods [135–137] were described and can be used to detect MYXV.

The conventional PCR system by Cavadini et al. [134] is based on the MYXV genes *M071L, M140R,* and *M142R/M144R*. Primers were designed for PCR and PCR-restriction fragment length polymorphism (PCR-RFLP) protocols, enabling

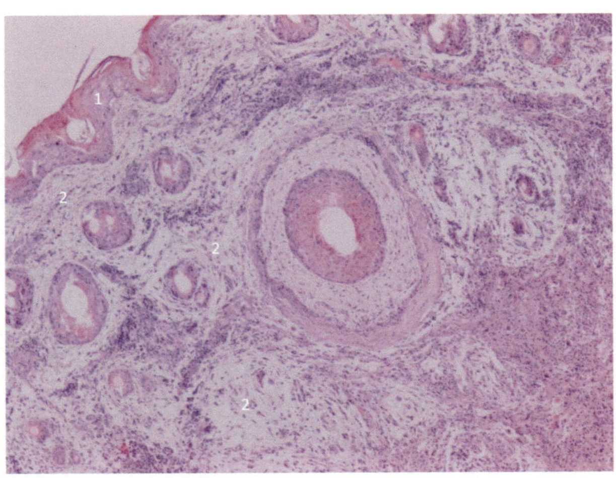


**Figure 10.**
*Histopathology of the lip of an Iberian hare with myxomatosis showing moderate hyperplasia of the epidermis (1) and myxoid tissue in the dermis (2). HE, 40× (photograph by the INIAV I.P. Pathology team).*

to discriminate vaccinated from naturally infected animals and to detect mixed infections caused by wild-type and vaccine MYXV strains.

Dalton et al. [138] developed a long-range PCR-RFLP method directed towards amplification of genomic MYXV DNA from the left and right terminal inverted repeat regions (TIRs) with subsequent differentiation of virus strains by RFLP. Two sets of primers were designed covering the entire TIRs and flanking sequences (*M009L* gene and genome regions from *M141R* to *M156R*). This method proved to be efficient in the identification of mutations, with potential application in phylogenetics.

Quantitative PCR systems (qPCR) allow the determination of viral copy number in the tested sample. In the PCR-based method developed by Albini et al. [136], primers and probe were designed to amplify a 147-bp fragment of the serpin (*Serp2*) gene, allowing the detection limit of 23 genome copies of MYXV DNA per reaction.

The TaqMan qPCR by Belsham et al. [135] that targets a *M029L* gene of MYXV was designed for detection and confirmation of suspected cases of myxomatosis. The assay detects efficiently as few as 10 copies of MYXV DNA, per reaction, while not producing amplification signals for other poxviruses.

A highly sensitive qPCR assay targeting nucleotide sequences within the MYXV gene *M000.5L/R* was developed by Duarte et al. [137]. This gene has two copies in the MYXV genome, in the right and left TIR, respectively. Hence, when compared to other PCR protocols targeting virus genes present in a single copy, this method shows a significantly higher sensitivity while enabling the detection of 2.6 genome copies of MYXV DNA per reaction. Furthermore, the *M000.5L/R* is a unique gene in the *Leporipoxvirus* genome [139] increasing the specificity of this PCR-based system. The method figures in the OIE Manual as a recommended method for nucleic acid detection (https://www.oie.int/fileadmin/Home/eng/Health_standards/tahm/3.06.01_MYXO.pdf).

The systems described by Cavadini et al. [134] and Duarte et al. [137] were successfully used to detect the recombinant MYXV obtained from hares (*our results unpublished*).

For many years, the laboratorial diagnosis of myxomatosis was based on the isolation of the virus in cell culture. Isolation of MYXV can be accomplished in primary cultures of rabbit kidney (RK) cells or in cell lines (RK-13, Vero). This method is recommended by the OIE, particularly for the diagnosis of amyxomatous form of the disease. In this case, antemortem diagnosis can be done using nasal and conjunctival swabs [140].

The virus multiplies in the cytoplasm of the infected cells [141, 142]. For the classical MYXV strains from rabbits, a cytopathic effect (CPE) may develop in 24–48 hours or take up to 1 week, depending on the virulence of the strain [143]. CPE is usually characterized by the formation of syncytia and roundup and contraction of the infected cells. Later, when all cells are affected, the cell monolayer detaches completely.

Viral isolation of the Iberian hare recombinant MYXV strains is more difficult since CPE is less obvious, differing from the one induced by the rabbit MYXV strains.

#### *2.2.3.5 Leporid herpesvirus*

Recently a new leporid herpesvirus (LeHV-5) was detected by PCR [144] and electron microscopy in Iberian hares [132].

Herpesvirus DNA was detected in hares with myxomatosis, where, in most cases, herpetic-like skin vesicles were present in the nostrils and lips along with necrosis of the genitalia, most evident in males affecting the penile glans but also observed in females (**Figure 11**) [132].

However, LeHV-5 DNA was also detected in apparently healthy hares, testing negative to MYXV, probably representing the latent stage of infection.

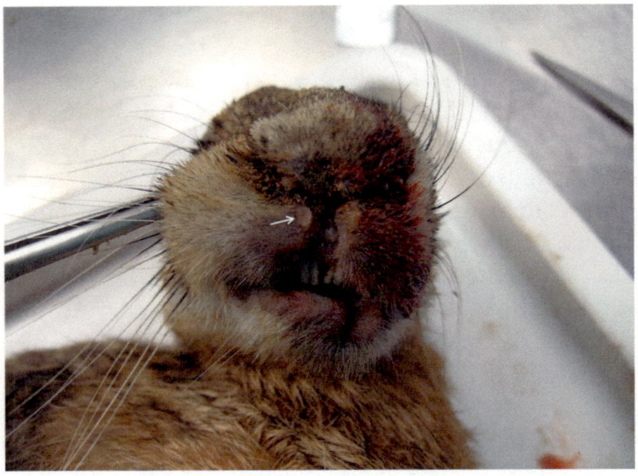

**Figure 11.**
*Small vesicle in the lips of an Iberian hare positive to herpesvirus (photograph by the INIAV I.P. Pathology team).*

The penile and foreskin epithelia of some LeHV-5 positive hares was mostly necrotic and replaced by a thick band of necrotic cells, heterophils, and red blood cells. Severe heterophilic infiltrations of the stroma, in either a diffuse pattern or multifocal aggregates, are common.

Proliferation of pleomorphic spindle cells is observed, with some nuclei almost completely filled with slightly eosinophilic inclusion bodies, resembling Cowdry nuclear inclusions [132]. Coalescent intraepidermal and subepidermal vesicopustules filled with fibrin and necrotic cell debris and multifocal detachment of the eyelids, lips, and foreskin epidermis were seen (**Figure 12**) [132]. In the dermis, multifocal hemorrhages, intense infiltration by heterophils, and necrotic cells with accumulation of chromatin debris were present [132].

Clinical diagnosis of this disease is difficult and may pass unnoticed. During lytic replication (reactivation phase), small vesicles may be observed in the mucous membranes and skin, which may still be intact, or after erosion, mainly in the lips

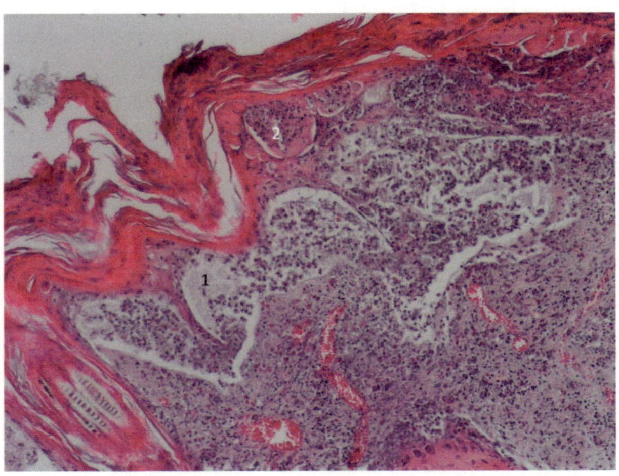


**Figure 12.**
*Subepidermal vesicular-pustular lesion with detachment of the epidermis (1) and intradermal vesicular-pustular lesion (2), from a herpesvirus positive Iberian hare. HE, 100× (photograph by the INIAV I.P. Pathology team).*

and nostrils. Necrosis of the genitalia may not be present. The conclusive diagnosis is achieved by PCR and sequencing or by electron microscopy.

### 2.3 Excessive predation

Predation can have a major impact on hare populations since, unlike the wild rabbit, hares generally do not burrow, except when subjected to high persecution pressure and without alternative escape, and are therefore more exposed to this phenomenon. The hares' natural predators include large birds of prey and wild canids and felids [29]. In fact, the Iberian hare is the preferred prey of some Iberian vertebrate predators, such as the Eurasian eagle owl (*Bubo bubo*) [145].

Some carnivorous bird species that can prey on hares have shown a progressive increase on the Iberian Peninsula. This is the case of the white stork (*Ciconia ciconia*) that between 1984 and 2004 showed a considerable increase in breeding numbers, as exposed by the number of occupied nests which increased about 401% in that period [146]. Furthermore, the number of wintering storks increased from 1187 individuals in 1995 to 14,434 in 2015 [147]. Large flocks of storks are presently found in many areas of mainland Portugal (**Figure 13**). The Eurasian or common magpie (*Pica pica*) is another omnivorous bird that may prey young hares [148].

The wild boar (*Sus scrofa*), a mammalian omnivore and one of the most widespread wildlife species which has entered a stage of continuous growth in Europe and could even be considered a pest species [149], can also have a potentially devastating impact on hare populations due to the easy predation of juvenile hares [150]. An increase in juvenile mortality reduces the recruitment of new individuals to populations, affecting their renewal.

### 2.4 Hunting pressure

Extrinsic factors threatening the conservation of hare populations include reduced size and quality of the habitat, which is particularly critical for hare species whose territories are restricted to very limited geographical areas, such as the northern Castroviejoi broom hare, in Spain, and the Italian hare, in Italy. In these small territories, excessive hunting pressure can pose a serious threat to the preservation of these species [5, 7].

In addition, the imbalance of the complex Iberian trophic chains resulting, among other aspects, from the reduction of wild rabbit populations, increases predation on hares by predators that preferentially feed on wild rabbit [31, 32].

**Figure 13.**
*Flock of Skorts, Alentejo, 15th March 2020 (Photograph by Fábio Abade dos Santos).*

### 2.5 Other causes

The mechanization of cereal harvesting was recognized as a cause of direct mortality of juvenile hares by their exposure on the soil surface. Also, the effect of pollution resulting from agricultural chemicals, agricultural and forest effluents [1], and road traffic pose additional threats to the species [10].

## 3. Sanitary surveillance of the Iberian hare in Portugal

In Portugal, an action plan to control rabbit viral hemorrhagic disease was set in place in August 2017 by a nine-institution partnership, following the Dispatch 4757/17 of May 31 of the Ministry of Agriculture. This partnership is comprised of the National Institute of Agrarian and Veterinary Research (INIAV. PI), the General Directorate of Food and Veterinary (DGAV), the Institute for Nature Conservation and Forests (ICNF), the Research Centre in Biodiversity and Genetic Resources (CIBIO), the Institute for Experimental Biology and Technology (iBET), the Portuguese first sector hunting organizations (FENCAÇA, ANPC, CNCP), and the National Order of Veterinary Doctors (OMV). Since its implementation from August 2017 to the end of January 2020, 1507 wild leporids originating from almost 50 hunting reserves where sampled and systematically tested for the presence of RHDV, RHDV2, and MYXV to assess the population health during the hunting seasons. Of these, 89.93% (1099/1222) were wild rabbits, and 10.07% (123/1222) were hares. Furthermore, during the same period, 285 animals found dead in the field in mainland Portugal were also screened, of which 77.54% (221/285) corresponded to wild rabbits and 22.46% (64/285) were hares.

Hunting associations are authorized by permits emitted by the National Forest Authority, Institute for Nature Conservation and Forests (ICNF).

Below we present the virological examinations regarding the Iberian hare. Despite RHDV2 being detected in several hare species, none of the 187 hares investigated in Portugal between August 2017 and the end of January 2020 were positive, confirming the lack of susceptibility of Iberian hare to this virus.

As in domestic and wild rabbits, where RHDV has no longer been detected in Portugal since the emergence of RHDV2 in 2012, RHDV-RNA was not detected in Iberian hares [151].

MYXV was first detected in Portugal in October 2018 in an adult hare, found dead in a hunting reserve, located in the district of Évora, Alentejo Region, in the south mainland. The virological diagnosis was made by the National Reference Laboratory for Animal Diseases (INIAV, Oeiras, Portugal) in October 2018.

The first evidence of Iberian hare mortality by myxomatosis was noticed in early summer of 2018, in Spain. The affected hares showed clinical signs compatible with myxomatosis and were found dead or moribund in hunting reserves in the municipalities of Montalbán and La Rambla (Córdoba), in the south. The diagnosis was confirmed by the Central Veterinary Laboratory in Algete, Madrid, Spain in July 2018 (OIE report).

Before the detection of the first MYXV case in the Iberian hare in Portugal, within the scope of the national surveillance action plan as referred above, 91 hares where sampled and screened for the presence of MYXV, between August 2017 and end of October 2018. None of the animals tested positive, suggesting that the virus was not circulating in the populations sampled. After the first case, by the end of October 2018 until the end of January 2020, 107 Iberian hares were tested for MYXV, 58.88% (63/107) of which were found dead or moribund and 41.12% (44/107) were legally hunted. In the first group, 82.54% (52/63) animals were positive for myxomatosis, reflecting the significance of this infection as a cause of death in Iberian hares.

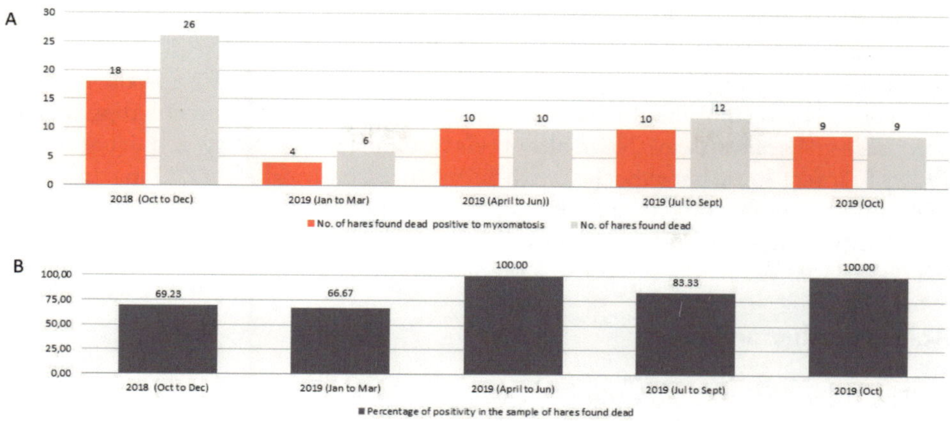


**Figure 14.**
*Results of the surveillance of myxoma virus in 63 Iberian hares found dead in continental Portugal between October 2018 and October 2019, by trimester (A). Percentage of positivity in the same sample (B).*

The percentage of positivity in the sample of hares found dead by trimester showed an increasing trend, suggesting the progressive spread of the disease (**Figure 14**).

In the group of 44 hunted animals, only 4 were positive (9.1%). Most of these samples ($n$ = 39, 86.4%) were collected during the 2018–2019 hunting season. The small sample gathered during the 2019–2020 hunting season ($n$ = 6, all collected in October 2019) does not allow to predict the trend of positivity for this group.

Since the non-recombinant rabbit MYXV is endemic in the entire national territory, our results suggest that the Iberian hare is not susceptible to the rabbit strains. However, serological data obtained from Iberian hares shows the presence of antibodies against RHDV2 and myxoma virus indicating the ability of these viruses to induce an immune response [151].

## 4. Conclusions

The progressive loss of habitat due to the deep changes in land use, excessive predation, and hunting poses serious threats to the conservation of hare species in Europe.

The development of effective strategies that trigger the continuous implementation of good practices is urgent to ensure the preservation of wild populations and to promote their recovery in the most affected areas. Such measures should safeguard the persistence/existence of favorable habitats for the species, particularly threatened in the geographical areas where their natural territories are reduced or in areas where farmers adopted super-intensive production methods.

In Portugal and in Spain, government research institutions, academics and field agents joined efforts through projects +Coelho and MIXolepus, respectively, to evaluate and counteract the effects of this emergent virus in Iberian hare.

At a time when the conservation status of the wild rabbit has recently been revised from "least concern" to "endangered," it seems inevitable that, in a near future, the status of some hare species, namely, the Iberian hare, will also be revised.

Apprehension on the preservation and sustainability of wild leporid populations in the Iberia is aligned with the concerns of many other wild species. According to the World Wildlife Fund (WWF) "Living Planet Report 2018" report, global wildlife populations have declined by an average of 60% over the past 40 years, demonstrating that the planetary biodiversity is threatened.

Infectious diseases, particularly those of viral etiology, have been gaining importance as disrupting factors for the stability of rabbits and hares. The potential evolution of the viral *hemorrhagic* disease virus (RHDV) to RHDV2 and the recent emergence of a recombinant myxoma virus, able to specifically infect hares, showed that these viruses were capable to alter their initial host specificity, acquiring the ability to infect some hare species, with very expressive morbidity and mortality rates. Furthermore, the recent detection of a new herpesvirus in Iberian hares, associated with genital pathology, raises addition concerns to the future of this species.

## Acknowledgements

The authors thank all the hunters who participated in the collection of biological samples, represented by João Carvalho (ANPC), Jacinto Amaro (FENCAÇA), and Fernando Castanheira Pinto (CNCP). Funding was supported by Fundo Florestal Permanente (Government of Portugal) within the scope of the action plan for the control of rabbit viral hemorrhagic disease (+Coelho 2 project nº 2018044300001 - Desenvolvimento e implementação de medidas práticas impulsionadoras da recuperação dos leporídeos silvestres em Portugal - financed by the Fundo Florestal Permanente (FFP), Dispatch nº4757/2017 of 31st May) by the Fundação para a Ciência e Tecnologia (FCT) (grant SFRH/BD/137067/2018) and by the Interdisciplinary Research Centre on Animal Health, Faculty of Veterinary Medicine, University of Lisbon (CIISA, FMV-UL) (Project UID/CVT/00276/2013).

## Conflict of interest

The authors declare no conflict of interest.

## Author details

Margarida D. Duarte[1,2*], Carina L. Carvalho[1], Fábio Abade dos Santos[1,2,3], Jéssica Monteiro[1], Madalena Monteiro[4], Paulo Melo Carvalho[4], Paula Mendonça[4], Patrícia Tavares Santos[5] and Pedro C. Melo[5]

1 National Institute of Agrarian and Veterinarian Research, Virology Laboratory, Oeiras, Portugal

2 Interdisciplinary Research Centre on Animal Health, Faculty of Veterinary Medicine, University of Lisbon (CIISA, FMV-UL), Portugal

3 Department of Biochemistry and Molecular Biology, University of Oviedo, Oviedo, Spain

4 National Institute of Agrarian and Veterinarian Research, Pathology Laboratory, Oeiras, Portugal

5 General Directorate of Food and Veterinary, Epidemiology and Animal Health Unit, Lisbon, Portugal

*Address all correspondence to: margarida.duarte@iniav.pt

## References

[1] Soriguer R, Carro F. *Lepus granatensis*. The IUCN Red List of Threatened Species 2019: e.T41306A29531952019. DOI: 10.2305/IUCN.UK.20191.RLTS. T41306A2953195.en (downloaded on 24 January 2020)

[2] Alves PC, Ferrand N, Suchentrunk F. Harris DJ, Ancient introgression of *Lepus timidus* mtDNA into *L. granatensis* and *L. europaeus* in the Iberian Peninsula. Molecular Phylogenetics and Evolution. 2003;**27**:70-80

[3] Smith AT, Johnston CH. *Lepus timidus*. The IUCN Red List of Threatened Species 2019: e.T11791A451771982019. DOI: 10.2305/IUCN.UK.2019-1.RLTS. T11791A45177198.en (downloaded on 24 January 2020)

[4] Paupério J. Ecologia de lebre-ibérica (*Lepus granatensis*) num Ecossistema de Montanha—Distribuição espacial, abundância e dieta de duas populações do Parque Natural da Serra da Estrela [Masters thesis]. Porto, Portugal: Faculdade de Ciências da Universidade do Porto; 2003

[5] Ballesteros F, Smith AT. *Lepus castroviejoi*. The IUCN Red List of Threatened Species 2019: e.T11797A5039082019. DOI: 10.2305/IUCN.UK.2019-2.RLTS.T11797A503908.en (downloaded on 24 January 2020)

[6] Hacklander K, Schai-Braun S. *Lepus europaeus*. The IUCN Red List of Threatened Species 2019: e.T41280A451874242019. DOI: 10.2305/IUCN.UK.2019-1.RLTS. T41280A45187424.en (downloaded on 24 January 2020)

[7] Randi E, Riga F. *Lepus corsicanus*. The IUCN Red List of Threatened Species 2019: e.T41305A29529542019. DOI: 10.2305/IUCN.UK.2019-2.RLTS. T41305A2952954.en (downloaded on 24 January 2020)

[8] Johnston CH, Robinson TJ, Child MF, Relton C. *Lepus capensis*. The IUCN Red List of Threatened Species 2019: e.T41277A451867502019. DOI: 10.2305/IUCN.UK.2019-1.RLTS.T41277A45186750.en (downloaded on 24 January 2020)

[9] Chapman JA, Flux JEC. Chapter 1: Introduction and overview of the lagomorphs. In: Chapman JA, Flux JEC, editors. Rabbits, Hares and Pikas. Status Survey and Conservation Action Plan. Gland, Switzerland: International Union for the Conservation of Nature and Natural Resources; 1990. pp. 1-13

[10] Purroy FJ. Liebre ibérica—*Lepus granatensis*. In: Salvador A, Cassinello J, editors. Enciclopedia Virtual de los Vertebrados Españoles. Madrid: Museo Nacional de Ciencias Naturales; 2011. Available from: http://www.vertebradosibericos.org/

[11] Smith AT, Johnston CH, Alves PC, Hackländer K, editors. Pikas, Rabbits and Hares of the World. Baltimore: Johns Hopkins University Press; 2018. LCCN 2017004268; ISBN: 9781421423401 (hardcover);ISBN: 1421423405 (hardcover); ISBN: 9781421423418 (electronic); ISBN: 1421423413 (electronic)

[12] de la Calzada E, Martínez FJ. Requerimientos y selección de hábitat de la liebre mediterrânea (*Lepus granatensis*) en un paisaje agrícola mesetario. Ecología. 1994;**8**:381-394

[13] López JMA, Hernández FJ, Purroy Y, Robles JL. Datos sobre la biología de la reproducción de la liebre Ibérica (*Lepus granatensis*) en agrosistemas cerealistas de la provincia de León (NW de España). Revista Florestal. 1996;**9**(1):49-60

[14] Duarte J. Liebre ibérica (*Lepus granatensis* Rosenhauer, 1856). Galemys. 2000;**12**:3-14

[15] Palacios F. Biometric and morphologic features of the species of the genus *Lepus* in Spain. Mammalia. 1989;**53**:227-263

[16] Marboutin E, Péroux YR. Some aspects of the spatial distribution of hares (*Lepus europaeus*) at night. Gibier Faune Sauvage. 1999;**16**(2):143-158

[17] Péroux R. Le lièvre d'Europe. Bulletim Mensusel de l'Office National la Chasse. 1995;**204**:1-96

[18] Chapman JA, Flux JEC, editors. Rabbits, Hares and Pikas. Status Survey and Conservation Action Plan. 1990. IUCN/SSC Lagomorph Specialist Group. IUCN/SSC Action Plans for the Conservation of Biological Diversity; Imprint: Gland: IUCN; 1990. ISBN: 2-8317-0019-1

[19] Holley AJF. The daily activity period of the brown hare (*Lepus europaeus*). Mammalian Biology. 2001;**66**:357-364

[20] Lovegrove B. Fires of Life. Endothermy in Birds and Mammals. New Haven, CT: Yale University Press; 2019

[21] JEC F. Reproductive strategies in the genus *Lepus*. In: Myers K, MacInnes CD, editors. Proceedings of the World Lagomorph Conference. Guelph, Ontario, Canada: University of Guelph; 1981. pp. 155-174

[22] Roellig K, Goeritz F, Fickel J, Hermes R, Hofer H, Hildebrandt TB. Superconception in mammalian pregnancy can be detected and increases reproductive output per breeding season. Nature Communications. 2010;**1**:78. DOI: 10.1038/ncomms1079

[23] Holley AJ, Greenwood PJ. The myth of the mad March hare. Nature (London). 1984;**309**:549-550

[24] Alves PC, Gonçalves H, Santos M, Rocha A. Original investigation reproductive biology of the Iberian hare, *Lepus granatensis*, in Portugal. Mammalian Biology. 2002;**67**:358-371. DOI: 10.1078/1616-5047-00051

[25] Parker SP. Grzimek's Encyclopedia of Mammals. New York: McGraw Hill Publishing Company; 1990

[26] Fernandez A, Soriguer R, Castien E, Carro F. Reproduction parameters of the Iberian hare *Lepus granatensis* at the edge of its range. Wildlife Biology. 2008;**14**:434-443. DOI: 10.2981/0909-6396-14.4.434

[27] Farfán MA, Vargas JM, Real R, Palomo LJ, Duarte J. Population parameters and reproductive biology of the Iberian hare Lepus granatensis in southern Iberia. Acta Theriologica. 2004;**49**(3):319-335

[28] Corbet GB. A review of classification in the family Leporidae. Acta Zoologica Fennica. 1983;**174**:11-15

[29] Taylor M. The Way of Hare. England: Bloomsbury Natural History, Bloomsbury Publishing Plc; 2017. p. 288. ISBN: 9781472909893

[30] BirdLife International 2019. *Aquila adalberti*. The IUCN Red List of Threatened Species 2019: e.T22696042A152593918. 10.2305/IUCN.UK.2019-3.RLTS.T22696042A152593918.en (downloaded on 12 February 2020)

[31] Gortazar C, Millán J, Acevedo P, Escudero MA, Marco J, de Luco DF. A large-scale survey of brown hare *Lepus europaeus* and Iberian hare *L. granatensis* populations at the limit of their ranges. Wildlife Biology. 2007;**13**:244-250. DOI: 10.2981/0909-6396(2007)13[244:ALSOBH]2.0.CO;2

[32] Acevedo P, Melo-Ferreira J, Real R, Alves PC. Past, present and future distributions of an Iberian endemic, *Lepus granatensis*: Ecological and evolutionary clues from species distribution models. PLoS One.

2012;**7**(12):e51529. DOI: 10.1371/journal.pone.0051529. [Epub: 13 December 2012]

[33] Francisco CARRO Ramón C. SORIGUER. Long-term patterns in Iberian hare population dynamics in a protected area (Doñana National Park) in the southwestern Iberian Peninsula: Effects of weather conditions and plant cover. Special subsection: Animal Behavior and Ecology. 2016. 10.1111/1749-4877.12212

[34] Marboutin E, Bray Y, Péroux R, Mauvy B, Lartiges A. Population dynamics in European hare: Breeding parameters and sustainable harvest rates. Journal of Applied Ecology. 2003;**40**:580-591. DOI: 10.1046/j.1365-2664.2003.00813.x

[35] Carro F, Beltrán JF, Pérez JM, Márquez FJ, Iborra O, Soriguer YRC. Dinámica Poblacional de la Liebre ibérica (*Lepus granatensis*) en el Parque Nacional de Doñana. Segovia: Resúmenes IV Jornadas SECEM; 1999. p. 20

[36] Voigt U, Siebert U. Living on the edge—Circadian habitat usage in pre-weaning European hares (*Lepus europaeus)* in an intensively used agricultural area. PLoS One. 2019;**14**(9):e0222205. DOI: 10.1371/journal.pone.0222205

[37] Duarte MD, Carvalho CL, Bernardo B, Barros SV, Benevides S, Flor L, et al. Rabbit haemorrhagic disease virus 2 (RHDV2) outbreak in Azores: Disclosure of common genetic markers and phylogenetic segregation within the European strains. Infection, Genetics and Evolution. 2015;**35**:163-171. DOI: 10.1016/j.meegid.2015.08.005. [Epub: 04 August 2015]

[38] Carvalho CL, Silva S, Gouveia P, Costa M, Duarte EL, Henriques AM, et al. Emergence of rabbit haemorrhagic disease virus 2 in the archipelago of Madeira, Portugal (2016-2017). Virus Genes. 2017;**53**(6):922-926. DOI: 10.1007/s11262-017-1483-6. [Epub 21 June 2017]

[39] Brown LJ, Rosatte RC, Fehlner-Gardiner C, Bachmann P, Ellison JA, Jackson FR, et al. Oral vaccination and protection of red foxes (*Vulpes vulpes*) against rabies using ONRAB, an adenovirus-rabies recombinant vaccine. Vaccine. 2014;**32**(8):984-989. DOI: 10.1016/j.vaccine.2013.12.015. [Epub: 24 December 2013]

[40] Maresch C, Lange E, Teifke JP, Fuchs W, Klupp B, Müller T, et al. Oral immunization of wild boar and domestic pigs with attenuated live vaccine protects against Pseudorabies virus infection. Veterinary Microbiology. 2012;**161**(1-2):20-25. DOI: 10.1016/j.vetmic.2012.07.002. [Epub: 10 July 2012]

[41] Wobeser G. Disease management strategies for wildlife. Revue Scientifique et Technique. 2002;**21**(1):159-178

[42] Bárcena J, Morales M, Vázquez B, Boga JA, Parra F, Lucientes J, et al. Horizontal transmissible protection against myxomatosis and rabbit hemorrhagic disease by using a recombinant myxoma virus. Journal of Virology. 2000;**74**:1114-1123

[43] Martín-Alonso JM, Castañón S, Alonso P, Parra F, Ordás R. Oral immunization using tuber extracts from transgenic potato plants expressing rabbit hemorrhagic disease virus capsid protein. Transgenic Research. 2003;**12**:127-130

[44] Parra F, Prieto M. Purification and characterization of a calicivirus as the causative agent of a lethal hemorrhagic disease in rabbits. Journal of Virology. 1990;**64**:4013-4015

[45] Carvalho CL. The role of wild leporids as reservoirs of infectious agents [PhD thesis]. Portugal: University of Évora; 2017. p. 431.

[46] Gavier-Widén D, Mörner T. Epidemiology and diagnosis of the European brown hare syndrome in Scandinavian countries. Revue Scientifique et Technique (International Office of Epizootics). 1991;**10**:453-458

[47] Wibbelt G, Frölich K. Infectious diseases in European Brown hare (*Lepus europaeus*). Wildlife Biology in Practice. 2005;**1**(1):86-93

[48] Lukešová D, Langrová I, Vadlejch J, Jankovská I, Hlava J, Válek P, et al. Endoparasites in European Hares (*Lepus europaeus*) Under Gamekeeping Conditions in the Czech Republic. Košice: Institute of Parasitology, SAS; 2012

[49] Diakou A, Sokos C, Papadopoulos E. Endoparasites Found in European Brown Hares (*Lepus europaeus*) Hunted in Macedonia, Greece. Research Note. Košice: Institute of Parasitology, SAS; 2014

[50] Kornaś S, Wierzbowska IA, Wajdzik M, Kowal J, Basiaga M, Nosal P. Endoparasites of European brown hare (*Lepus europaeus*) from southern Poland based on necropsy. Annals of Animal Science. 2014;**14**(2):297-305

[51] Carvalho CL, Zé-Zé L, Lopes de Carvalho I, Duarte EL. Tularaemia: a challenging zoonosis. Comparative Immunology, Microbiology and Infectious Diseases. 2014;**37**(2):85-96. DOI: 10.1016/j.cimid.2014.01.002. [Epub: 13 January 2014]

[52] Ruiz-Fons F, Ferroglio E, Gortázar C. *Leishmania infantum* in free-ranging hares, Spain, 2004-2010. Eurosurveillance. 2013;**18**(30):20541

[53] Lopes de Carvalho I, Toledo A, Carvalho CL, Barandika JF, Respicio-Kingry LB, Garcia-Amil C, et al. *Francisella* species in ticks and animals, Iberian Peninsula. Ticks Tick Borne Diseases. 2016;**7**(1):159-165. DOI: 10.1016/j.ttbdis.2015.10.009. [Epub: 19 October 2015]

[54] Martín C, Gallardo MT, Mateos L, Vián E, García MJ, Ramos J, et al. Outbreak of tularaemia in Castilla y León, Spain. Eurosurveillance. 2007;**12**(11):E071108.1. DOI: 10.2807/esw.12.45.03302-en

[55] Gyuranecz M, Erdélyi K, Makrai L, Fodor L, Szépe B, Mészáros AR, et al. Brucellosis of the European brown hare (*Lepus europaeus*). Journal of Comparative Pathology. 2011;**145**(1):1-5. DOI: 10.1016/j.jcpa.2010.11.013

[56] Schantz PM, Cruz-Reyes A, Colli C, Lord RD. Sylvatic echinococcosis in Argentina. I. On the morphology and biology of strobilar *Echinococcus granulosus* (Batsch, 1786) from domestic and sylvatic animal hosts. Tropenmedizin und Parasitologie. 1975;**26**(3):334-344

[57] Craig P, Mastin A, van Kesteren F, Boufana B. *Echinococcus granulosus*: Epidemiology and state-of-the-art of diagnostics in animals. Veterinary Parasitology. 2015;**213**(3-4):132-148. DOI: 10.1016/j.vetpar.2015.07.028

[58] Chaignat V, Boujon P, Frey CF, Hentrich B, Müller N, Gottstein B. The brown hare (*Lepus europaeus*) as a novel intermediate host for *Echinococcus multilocularis* in Europe. Parasitology Research. 2015;**114**(8):3167-3169. DOI: 10.1007/s00436-015-4555-3

[59] Magistrali CF, Cucco L, Pezzotti G, Farneti S, Cambiotti V, Catania S, et al. Characterisation of *Yersinia pseudotuberculosis* isolated from animals with yersiniosis during 1996-2013 indicates the presence of pathogenic and Far Eastern strains in Italy. Veterinary Microbiology, 166. 2015;**180**:161. DOI: 10.1016/j.vetmic.2015.08.020

[60] Fratini F, Verin R, Ebani VV, Ambrogi C, Bertelloni F, Turchi B, et al. Experimental infection with *Yersinia pseudotuberculosis* in European brown hare (*Lepus europaeus*, Pallas). Asian Pacific Journal of Tropical Medicine.

2017;**10**(3):285-291. DOI: 10.1016/j.apjtm.2017.03.008

[61] Stancampiano L, Ravagnan S, Capelli G, Militerno G. Cysticercosis by *Taenia pisiformis* in Brown hare (*Lepus europaeus*) in Northern Italy: Epidemiologic and pathologic features. The International Journal for Parasitology: Parasites and Wildlife. 2019;**9**:139-143. DOI: 10.1016/j.ijppaw.2019.04.004

[62] Camarda A, Pugliese N, Cavadini P, Circella E, Capucci L, Caroli A, et al. Detection of the new emerging rabbit haemorrhagic disease type 2 virus (RHDV2) in Sicily from rabbit (*Oryctolagus cuniculus*) and Italian hare (*Lepus corsicanus*). Research in Veterinary Science. 2014;**97**:642-645. DOI: 10.1016/j.rvsc.2014.10.008

[63] Puggioni G, Cavadini P, Maestrale C, Scivoli R, Botti G, Ligios C, et al. The new French 2010 rabbit hemorrhagic disease virus causes an RHD-like disease in the Sardinian Cape hare (*Lepus capensis mediterraneus*). Veterinary Research. 2013;**44**:96. DOI: 10.1186/1297-9716-44-96

[64] Le Gall-Reculé G, Lemaitre E, Bertagnoli S, Hubert C, Top S, Decors A, et al. Large-scale lagovirus disease outbreaks in European brown hares (*Lepus europaeus*) in France caused by RHDV2 strains spatially shared with rabbits (*Oryctolagus cuniculus*). Veterinary Research. 2017;**48**(1):70. DOI: 10.1186/s13567-017-0473-y

[65] Velarde R, Cavadini P, Neimanis A, Cabezón O, Chiari M, Gaffuri A, et al. Spillover events of infection of brown hares (*Lepus europaeus*) with rabbit haemorrhagic disease type 2 virus (RHDV2) caused sporadic cases of an European Brown Hare Syndrome-*like* disease in Italy and Spain. Transboundary and Emerging Diseases. 2017;**64**(6):1750-1761. DOI: 10.1111/tbed.12562

[66] Neimanis AS, Ahola H, Larsson Pettersson U, Lopes AM, Abrantes J, Zohari S, et al. Overcoming species barriers: An outbreak of Lagovirus europaeus GI.2/RHDV2 in an isolated population of mountain hares (*Lepus timidus*). BMC Veterinary Research. 2018;**14**(1):367. DOI: 10.1186/s12917-018-1694-7

[67] Bell DJ, Davis JP, Gardner M, Barlow AM, Rocchi M, Gentil M, Wilson RJ. Rabbit haemorrhagic disease virus type 2 in hares in England. The Veterinary Record. 2019. Jan 26;184(4):127-128. DOI: 10.1136/vr.l337.

[68] García-Bocanegra I, Camacho-Sillero L, Risalde MA, Dalton KP, Caballero-Gómez J, Agüero M, et al. First outbreak of myxomatosis in Iberian hares (*Lepus granatensis*). Transboundary and Emerging Diseases. 2019;**66**(6):2204-2208. DOI: 10.1111/tbed.13289

[69] Águeda-Pinto A, Lemos de Matos A, Abrantes M, Kraberger S, Risalde MA, Gortázar C, et al. Genetic characterization of a recombinant myxoma virus in the Iberian hare (*Lepus granatensis*). Viruses. 2019;**11**(6). pii: E530. DOI: 10.3390/v11060530

[70] Dalton KP, Martín JM, Nicieza I, Podadera A, de Llano D, Casais R, et al. Myxoma virus jumps species to the Iberian hare. Transboundary and Emerging Diseases. 2019;**66**(6):2218-2226. DOI: 10.1111/tbed.13296

[71] Jones A, Pybus MJ. Taeniasis and echinococcosis. In: Samuel WM, Pybus MJ, Kocan AA, editors. Parasitic Diseases of Wild Mammals. Ames, USA: Iowa State University Press; 2001. pp. 150-192

[72] Contreiro J. Prevalência de Taenia pisiformis/Cysticercus pisiformis em três zonas de caça do Baixo Alentejo [MSc thesis]. Universidade Lusófona de Humanidades e Tecnologias Faculdade de Medicina Veterinária Lisboa; 2014

[73] Helminths SEJL. Helminths, Arthropods and Protozoa of Domestic Animals. Baltimore: Williams and Wilkins; 1968

[74] Graham-Brown J, Gilmore P, Harcourt-Brown F, Eastham H, Williams D. Veterinary Record Case Reports. 2018. Published Online First. DOI: 10.1136/vetreccr-2018-000634

[75] Hallal-Calleros C, Morales-Montor J, Orihuela-Trujillo A, Togno-Peirce C, Murcia-Mejía C, Bielli A, et al. *Taenia pisiformis* cysticercosis induces decreased prolificacy and increased progesterone levels in rabbits. Veterinary Parasitology. 2016;**229**:50-53. DOI: 10.1016/j.vetpar.2016.09.015

[76] Alzaga V, Vicente J, Villanua D, Acevedo P, Casas F, Gortazar C. Body condition and parasite intensity correlates with escape capacity in Iberian hares (*Lepus granatensis*). Behavioral Ecology and Sociobiology. 2008;**62**:769-775

[77] Harcourt-Brown FM. Infectious diseases of domestic rabbits. In: Textbook of Rabbit Medicine. 1st ed. Oxford: Butterworth-Heinemann; 2001. p. 363

[78] Saunders RA, Rees-Davies R. Notes on Rabbit Internal Medicine. Oxford: Blackwell; 2005. pp. 146-147

[79] Varga M. The rabbit-friendly practice. In: Meredith A, Lord B, editors. BSAVA Manual of Rabbit Medicine. Gloucester: British Small Animal Veterinary Association; 2014. p. 66

[80] Manning PJ, Ringler DH, Newcomer CE. The Biology of the Laboratory Rabbit. 2nd ed. London: Academic Press Limited. ISBN: 0-12-469235-4

[81] Van Praag, E. Cysticercosis and Hydatid Echinococcosis, Dangerous Parasitic Larval Forms in Rabbits. 2015

[82] Shield JM, Heath DD, Smyth JD. Light microscope studies of the early development of *Taenia pisiformis* cysticerci. International Journal for Parasitology. 1973;**3**(4):471-480. DOI: 10.1016/0020-7519(73)90042-8

[83] Khalil A, Noor El Din S, Radwan N, et al. Cysticercus pisiformis: Ultrastructural transformation of the tegument during development from oncosphere to cysticercus. Parasitologists United Journal. 2014;**7**:13

[84] Solomon SG. Some points in the early development of cysticercus pisiformis (Bloch 1780). Journal of Helminthology. 1934;**12**:197-204

[85] Heath DD. The migration of oncospheres of *Taenia pisiformis*, *T. serialis* and *Echinococcus granulosus* within the intermediate host. International Journal for Parasitology. 1971;**1**:145-152. DOI: 10.1016/0020-7519(71)90008-7

[86] Boubaker G, Marinova I, Gori F, Hizem A, Müller N, Casulli A, et al. A dual PCR-based sequencing approach for the identification and discrimination of *Echinococcus* and *Taenia* taxa. Molecular and Cellular Probes. 2016;**30**(4):211-217. DOI: 10.1016/j.mcp.2016.05.004

[87] Duff JP, Whitwell K, Chase D. The emergence and epidemiology of European brown hare syndrome in the UK. In: Proc, 1st International Symposium on Caliciviruses ESVV, Reading, UK. 1997. pp. 176-181

[88] Lenghaus C, Studdert MJ, Gavier-Widden D. Calicivirus infections. In: Williams ES, Barker IK, editors. Infectious Disease of Wild Mammals. 3rd ed. Ames, Iowa: Iowa State University Press; 2001. pp. 280-291

[89] Scicluna MT, Lavazza A, Capucci L. European brown hare syndrome in northern Italy: Results of a virological and serological survey. Revue Scientifique et Technique.

1994;**13**(3):893-894. DOI: 10.20506/rst.13.3.801

[90] Chasey D, Lucas M, Westcott D, Williams M. European brown hare syndrome in the UK: A calicivirus related to but distinct from that of viral haemorrhagic disease in rabbits. Archives of Virology. 1992;**124**(3-4): 363-370. DOI: 10.1007/bf01309816

[91] Le Gall-Reculé G, Zwingelstein F, Laurent S, Portejoie Y, Rasschaert D. Molecular epidemiology of European brown hare syndrome virus in France between 1989 and 2003. Archives of Virology. 2006;**151**(9):1713-1721. DOI: 10.1007/s00705-006-0754-7. [Epub: 07 April 2006]

[92] Frölich K, Meyer HH, Pielowski Z, Ronsholt L, von Seck-Lanzendorf S, Stolte M. European brown hare syndrome in free-ranging hares in Poland. Journal of Wildlife Diseases. 1996;**32**(2):280-285. DOI: 10.7589/0090-3558-32.2.280

[93] Billinis C, Knowles NJ, Spyrou V, Sofianidis G, Psychas V, Birtsas PK, et al. Genetic analysis of the first European Brown hare syndrome virus isolates from Greece. Wildlife Biology in Practice. 2005;**1**(2):118-127. DOI: 10.2461/wbp.2005.1.14

[94] Frölich K, Fickel J, Ludwig A, Lieckfeldt D, Streich WJ, Jurcik R, et al. New variants of European brown hare syndrome virus strains in free-ranging European brown hares (*Lepus europaeus*) from Slovakia. Journal of Wildlife Diseases. 2007;**43**(1):89-96

[95] Syrjälä P, Nylund M, Heinikainen S. European brown hare syndrome in free-living mountain hares (*Lepus timidus*) and European brown hares (*Lepus europaeus*) in Finland 1990-2002. Journal of Wildlife Diseases. 2005;**41**(1):42-47. DOI: 10.7589/0090-3558-41.1.42

[96] Capucci L, Scicluna MT, Lavazza A. Diagnosis of viral haemorrhagic disease of rabbits and the European brown hare syndrome. Revue Scientifique et Technique. 1991;**10**:347-370

[97] Fuchs A, Weissenböck H. Comparative histopathological study of rabbit haemorrhagic disease (RHD) and European brown hare syndrome (EBHS). Journal of Comparative Pathology. 1992;**107**(1):103-113

[98] Wirblich C, Meyers G, Ohlinger VF, Capucci L, Eskens U, Haas B, et al. European brown hare syndrome virus: Relationship to rabbit hemorrhagic disease virus and other caliciviruses. Journal of Virology. 1994;**68**(8):5164-5173

[99] Ros Bascuñana C, Nowotny N, Belák S. Detection and differentiation of rabbit hemorrhagic disease and European brown hare syndrome viruses by amplification of VP60 genomic sequences from fresh and fixed tissue specimens. Journal of Clinical Microbiology. 1997;**35**(10):2492-2495

[100] Duff JP, Chasey D, Munro R, Wooldridge M. European brown hare syndrome in England. The Veterinary Record. 1994;**134**(26):669-673. DOI: 10.1136/vr.134.26.669

[101] Le Gall-Reculé G, Zwingelstein F, Portejoie Y, Le Gall G. Immunocapture-RT-PCR assay for detection and molecular epidemiology studies of rabbit haemorrhagic disease and European Brown Hare Syndrome viruses. Journal of Virological Methods. 2001;**97**(1-2):49-57. DOI: 10.1016/s0166-0934(01)00336-6

[102] Zexiao Y, Lirui L, Yin W, Xueping Y, Xueqing H, Zengqi Y, et al. Development of a RT-PCR for the detection of European brown hare syndrome virus. Advanced Materials Research. 2011;**271-273**:410-416

[103] Liu SJ, Xue HP, Pu BQ, Qian NH. A new viral disease of rabbits. Animal Husbandry & Veterinary Medicine. 1984;**16**:253-255

[104] Le Gall-Reculé G, Zwingelstein F, Boucher S, Le Normand B, Plassiart G, Portejoie Y, et al. Detection of a new variant of rabbit haemorrhagic disease virus in France. The Veterinary Record. 2011;**168**:137-138. DOI: 10.1136/vr.d697

[105] Lopes AM, Marques S, Silva E, Magalhães MJ, Pinheiro A, Alves PC, et al. Detection of RHDV strains in the Iberian hare (*Lepus granatensis*): Earliest evidence of rabbit lagovirus cross-species infection. Veterinary Research. 2014;**45**:94. DOI: 10.1186/s13567-014-0094-7

[106] Hall RN, Peacock DE, Kovaliski J, Mahar JE, Mourant R, Piper M, et al. Detection of RHDV2 in European brown hares (*Lepus europaeus*) in Australia. The Veterinary Record. 2017;**180**(5):121. DOI: 10.1136/vr.104034

[107] Rocchi M, Maley M, Dagleish M, Boag B. Rabbit haemorrhagic disease virus type 2 in hares in Scotland. The Veterinary Record. 2019;**185**(1):23. DOI: 10.1136/vr.l4481

[108] World Organisation for Animal Health (OIE) Technical Disease Cards. Rabbit Haemorrhagic Disease. Aetology, Epidemiology, Diagnosis, Prevention and Control References. 2015. Available from: https://www.oie.int/fileadmin/Home/eng/Animal_Health_in_the_World/docs/pdf/Disease_cards/RHD.pdf

[109] Le Gall-Reculé G, Lavazza A, Marchandeau S, Bertagnoli S, Zwingelstein F, Cavadini P, et al. Emergence of a new lagovirus related to rabbit haemorrhagic disease virus. Veterinary Research. 2013;**44**:81. DOI: 10.1186/1297-9716-44-81

[110] Dalton KP, Nicieza I, Balseiro A, Muguerza MA, Rosell JM, Casais R, et al. Variant rabbit hemorrhagic disease virus in young rabbits, Spain. Emerging Infectious Diseases. 2012;**18**:2009-2012. DOI: 10.3201/eid1812.120341

[111] Guittré C, Baginski I, Le Gall G, Prave M, Trepo C, Cova L. Detection of rabbit haemorrhagic disease virus isolates and sequence comparison of the N-terminus of the capsid protein gene by the polymerase chain reaction. Research in Veterinary Science. 1995;**58**:128-132. DOI: 10.1016/0034-5288(95)90065-9

[112] Tham KM, Barnes SM, Hunter SN. Polymerase chain reaction amplification and gene sequence analysis of a calicivirus from a feral rabbit. Virus Genes. 1999;**18**(3):235-242. DOI: 10.1023/a:1008020303036

[113] Yang L, Wang F, Hu B, Xue J, Hu Y, Zhou B, et al. Development of an RT-PCR for rabbit hemorrhagic disease virus (RHDV) and the epidemiology of RHDV in three eastern provinces of China. Journal of Virological Methods. 2008;**151**:24-29. DOI: 10.1016/j.jviromet.2008.04.003

[114] Gall A, Hoffmann B, Teifke JP, Lange B, Schirrmeier H. Persistence of viral RNA in rabbits which overcome an experimental RHDV infection detected by a highly sensitive multiplex real-time RT-PCR. Veterinary Microbiology. 2007;**120**(1-2):17-32. DOI: 10.1016/j.vetmic.2006.10.006

[115] Yuan D, Guo D, Liu J, Si C, Jiang Q, Lin H, et al. Development of a reverse-transcription loop-mediated isothermal amplification method for detection of rabbit hemorrhagic disease virus. Journal of Virological Methods. 2013;**187**:274-277. DOI: 10.1016/j.jviromet.2012.11.020

[116] Niedzwiedzka-Rystwej P, Hukowska-Szematowicz B, Dzialo J, Tokarz-Deptula B, Deptula W. Real time PCR detection of rabbit haemorrhagic disease virus in rabbits infected with different European strains of RHDV. Polish Journal of Veterinary Sciences. 2013;**16**:39-43. DOI: 10.2478/pjvs-2013-0006

[117] Duarte MD, Carvalho CL, Barros SC, Henriques AM, Ramos F, Fagulha T, et al. A real time Taqman RT-PCR for the detection of rabbit hemorrhagic disease virus 2 (RHDV2). Journal of Virological Methods. 2015;**219**:90-95. DOI: 10.1016/j.jviromet.2015.03.017

[118] Carvalho CL, Duarte EL, Monteiro M, Botelho A, Albuquerque T, Fevereiro M, et al. Challenges in the rabbit haemorrhagic disease 2 (RHDV2) molecular diagnosis of vaccinated rabbits. Veterinary Microbiology. 2017;**98**:43-50. DOI: 10.1016/j.vetmic.2016.12.006

[119] Carvalho CL, Duarte EL, Monteiro JM, Afonso C, Pacheco J, Carvalho P, et al. Progression of rabbit haemorrhagic disease virus 2 upon vaccination in an industrial rabbitry: A laboratorial approach. World Rabbit Science. 2017;**25**:73-85

[120] Carvalho CL, Abade dos Santos F, Fagulha T, Duarte MD. The aptitude of the specific RT-qPCR method used for RHDV2 detection: *In silico* analysis. In: Poster. Wildlife Summit—Game Manegement, 28 and 29 of June, 2019. Oeiras, Portugal

[121] Dalton KP, Arnal JL, Benito AA, Chacón G, Martín Alonso JM, Parra F. Conventional and real time RT-PCR assays for the detection and differentiation of variant rabbit hemorrhagic disease virus (RHDVb) and its recombinants. Journal of Virological Methods. January 2018;**251**:118-122. DOI: 10.1016/j.jviromet.2017.10.009

[122] Lucas A, Bouley G, Quinchon C, Tocas L. La myxomatose du lievre. Bulletin de l'Office International des Epizooties. 1953;**39**:770-776

[123] Jacotot H, Vallee A, Virat B. Sur un cas de myxomatose chez le lievre. Annales del'Institut Pasteur. 1954;**86**:105-107

[124] Barlow A, Lawrence K, Everest D, Dastjerdi A, Finnegan C, Steinbach F. Confirmation of myxomatosis in a European brown hare in Great Britain. The Veterinary Record. 2014;**175**(3):75-76. DOI: 10.1136/vr.g4621

[125] OIE. Myxomatosis. Spain. 2018. Available from: https://www.oie.int/wahis_2/public/wahid.php/Reviewreport/Review/viewsummary?fupser=&dothis=&reportid=27539

[126] Carvalho CL, Abade dos Santos FA, Monteiro M, Carvalho P, Mendonça P, Duarte MD. First cases of myxomatosis in Iberian hares (*Lepus granatensis*) Veterinary Record Case Reports (Portugal). 2020. DOI: 10.1136/vetreccr-2019-001044 [Accessed: 13 April 2020]

[127] Stanford MM, Werden SJ, McFadden G. Myxoma virus in the European rabbit: Interactions between the virus and its susceptible host. Veterinary Research. 2007;**38**(2): 299-318. DOI: 10.1051/vetres:2006054. [Epub: 13 February 2007]

[128] Kerr PJ, Liu J, Cattadori I, Ghedin E, Read AF, Holmes EC. Myxoma virus and the Leporipoxviruses: An evolutionary paradigm. Viruses. 2015;**7**(3):1020-1061. DOI: 10.3390/v7031020

[129] Bertagnoli S, Marchandeau S. Myxomatosis. Revue Scientifique et Technique (International Office of Epizootics). 2015;**34**(2):549-556

[130] Marlier D, Vindevogel H. Poxless myxomatosis—Isolation of three strains in Belgium. Annales de Médecine Vétérinaire. 1996;**140**:343-346

[131] Marlier D, Mainil J, Linde A, Vindevogel H. Infectious agents associated with rabbit pneumonia: Isolation of amyxomatous myxoma virus strains. The Veterinary Journal.

2000;**159**(2):171-178. DOI: 10.1053/tvjl.1999.0413

[132] Abade dos Santos FA, Monteiro M, Pinto A, Carvalho CL, Peleteiro MC, Carvalho P, et al. First description of a herpesvirus infection in genus Lepus. PLoS One. 2020;**14**:1-20. DOI: 10.1371/journal.pone.0231795

[133] Dinev I. An outbreak of myxomatosis in rabbits in bulgaria clinicomorphological studies. Trakia Journal of Sciences. 2012;**10**(1):79-84. Available from: http://www.uni-sz.bg

[134] Cavadini P, Botti G, Barbieri I, Lavazza A, Capucci L. Molecular characterization of SG33 and Borghi vaccines used against myxomatosis. Vaccine. 2010;**28**:5414-5420. DOI: 10.1016/j.vaccine.2010.06.017

[135] Belsham GJ, Polacek C, Breum SØ, Larsen LE, Bøtner A. Detection of myxoma viruses encoding a defective M135R gene from clinical cases of myxomatosis; possible implications for the role of the M135R protein as a virulence factor. Virology Journal. 2010;7:7-17. DOI: 10.1186/1743-422X-7-7

[136] Albini S, Sigrist B, Güttinger R, Schelling C, Hoop RK, Vögtlin A. Development and validation of myxoma virus real-time polymerase chain reaction assay. Journal of Veterinary Diagnostic Investigation. 2012;**24**:135-137. DOI: 10.1177/1040638711425946

[137] Duarte MD, Barros SC, Henriques AM, Fagulha MT, Ramos F, Luís T, et al. Development and validation of a real time PCR for the detection of myxoma virus based on the diploid gene M000.5L/R. The Journal of Virological Methods. 2014;**196**:219-224. DOI: 10.1016/j.jviromet.2013.11.014

[138] Dalton KP, Ringleb F, Martín Alonso JM, Parra F. Rapid identification of myxoma virus variants by long-range PCR and restriction fragment length polymorphism analysis. Journal of Virological Methods. 2009;**161**:284-288. DOI: 10.1016/j.jviromet.2009.06.026

[139] Willer DO, McFadden G, Evans DH. The complete genome sequence of Shope (rabbit) fibroma virus. Virology. 1999;**264**:319-343. DOI: 10.1006/viro.1999.0002

[140] OIE (World Organisation for Animal Health) Terrestrial Manual 2018. Chapter 3.6.1. Myxomatosis. Available from: http://www.oie.int/en/international-standardsetting/terrestrial-manual/access-online/

[141] Murphy FA, Fauquet CM, Bishop DHL, Ghabrial SA, Jarvis AW, Martelli GP, et al. Virus taxonomy: classification and nomenclature of viruses. In: Sixth Report of the International Committee for the Taxonomy of Viruses. Archives of Virology. Vienna: Springer Verlag; 1995 (Supplement 10). pp. 586

[142] Scherrer PR. Morphogenese *et* ultrastructure de virus fibromateux de Shope. Pathology and Microbiology. 1968;**3**:129-146

[143] Joubert L. La Myxomatose T.II. Série: Les Maladies Animales à Virus. Paris, France: L'Expansion éditeur; 1973

[144] VanDevanter DR, Warrener P, Bennett L, et al. Detection and analysis of diverse herpesviral species by consensus primer PCR. Journal of Clinical Microbiology. 1996;**34**(7):1666-1671

[145] Lourenço RF. The food habits of Eurasian eagle-owls in Southern Portugal. The Journal of Raptor Research. 2006;**40**:297-300

[146] Rosa G, Encarnação V. The white stork in Portugal: Results of the 2004 National Census. Lisbon, Portugal:

NABU-Bundesverband; 2013. Available from: https://bergenhusen.nabu.de/imperia/md/nabu/images/nabu/einrichtungen/bergenhusen/portugal_zensus_2004.pdf

[147] Catry I, Encarnação V, Pacheco C, Catry T, Tenreiro P, da Silva LP, et al. Recent changes on migratory behaviour of the White stork (*Ciconia ciconia*) in Portugal: Towards the end of migration? Airo. 2016/2017;**24**:28-35

[148] Jennings N. RSPB Spotlight Hares. England: Bloomsbury Natural History, Bloomsbury Publishing Plc; 2017. pp. 128. ISBN-10: 1472933648; ISBN-13: 978-1472933645

[149] Veličković N, Ferreira E, Djan M, Ernst M, Obreht Vidaković D, Monaco A, et al. Demographic history, current expansion and future management challenges of wild boar populations in the Balkans and Europe. Heredity (Edinb). 2016;**117, 5**:348-357. DOI: 10.1038/hdy.2016.53

[150] Schley L, Roper TJ. Diet of wild boar *Sus scrofa* in Western Europe, with particular reference to consumption of agricultural crops. Mammal Review. 2003;**33**(1):43-56 (Printed in Great Britain)

[151] Duarte M, Carvalho CL, Santos FA, Gomes J, Alves PC, Esteves PJ, et al. "+Coelho: Avaliação Ecossanitária das Populações Naturais de Coelho Bravo Visando o Controlo da Doença Hemorrágica Viral". Relatório de atividades na época venatória 2017-2018. INIAV I.P., DGAV, CIBIO, ANPC, CNCP, FENCAÇA. Fundo Florestal Permanente. ICNF. 2018. pp. 280

Chapter 2

# Viral Disease in Lagomorphs: A Molecular Perspective

*Kevin P. Dalton, Ana Podadera, José Manuel Martin Alonso, Inés Calonge Sanz, Ángel Luis Álvarez Rodríguez, Rosa Casais and Francisco Parra*

## Abstract

Our understanding of molecular biology of the viruses that infect lagomorphs is largely limited to the leporipoxvirus myxoma virus (MYXV) and the lagoviruses rabbit haemorrhagic disease virus (RHDV) and European brown hare syndrome virus (EBHSV) that infect the European rabbit (*Oryctolagus cuniculus*) and the European brown hare (*Lepus europaeus*) respectively. Thanks to the great effort of historic surveillance studies and careful sample archiving, the molecular evolution of these viruses is being resolved. Although historically considered viruses that cause species specific diseases recent reports show that several lagomorphs may now face the threat of these maladies. The driving factors behind these changes has not been determined and the effect of these species jumps on lagomorph populations has yet to be seen. Lagomorphs are also affected by several other lesser studied viral diseases. In addition, recent metagenomic studies have led to the identification of novel lagomorph viruses the importance of these to lagomorph health remains to be fully determined. In this chapter we summarize molecular aspects of viruses that infect lagomorphs, paying particular attention to recent interspecies infections.

**Keywords:** myxoma virus, rabbit hemorrhagic disease virus, lagomorphs, molecular virology, rabbit, hare, species jump

## 1. Introduction

Two viral diseases affecting Leporidae are listed in the World Organization for Animal Health Terrestrial Manual for animals, myxomatosis and rabbit hemorrhagic disease. Both are of major significance to wild and domestic rabbits and hares causing important environmental harm with significant financial consequences. This is also reflected in the magnitude of scientific literature regarding these diseases and the viruses that cause them namely myxoma virus (MYXV) and rabbit hemorrhagic disease virus (RHDV). MYXV and RHDV belong to two virus families the *Poxviridae* and *Caliciviridae* respectively. These virus families include a number of other viruses that are of consequence to lagomorphs. In the first section of this chapter we consider the molecular biology of viruses from these two virus families. Lagomorphs are also subject to a broader range of viral infections that we shall outline in part II of the chapter.

## 2. Viruses that infect lagomorphs Part I

### 2.1 Leporipoxvirus (virus family Poxviridae)

#### *2.1.1 A brief history; species jump and evolving virulence*

There are currently two *leporipoxvirus* (virus family *Poxviridae*) of consequence for lagomorphs, myxoma virus (MYXV) and rabbit fibroma virus (RFV). Both cause tumorifications in their natural hosts *Silvilagus brasiliensis* (the Brazilian cottontail) and *Sylvilagus floridanus* (the eastern cottontail rabbit), respectively. However, they lead to considerably different outcomes following infection of the European rabbit (*Oryctolagus cuniculus*). RFV may cause myxomas in the European rabbit but animals mount an immune response and recover, MYXV however causes lethal myxomatosis.

Myxomatosis was first observed by Sanarelli [1] in European rabbits (*Oryctolagus cuniculus*) imported to Uruguay. Local rabbits (*Silvilagus brasiliensis*) (tapeti) were the source of the virus that caused this devastating disease. The disease is one of the best studied examples of a virus species jump or cross-species transmission. The emergence of myxomatosis in the European rabbit in the broader context is directly linked to human activity through deliberate releases of infected animals on three continents. The aim was to control populations of rabbits considered pests in Australia, France and Chile. (reviewed in [2–5]). Seventy years since its introduction the disease continues to persist. Through extensive monitoring programs and comprehensive virulence studies an evolutionary model of host-pathogen interactions was developed (reviewed in [5]).

Following the introduction of the disease in Australia attenuated strains of the virus soon arose [4] this was accompanied by the selection of more resistant rabbits. The comparison of rabbit genomes from before and after the introduction of MYXV has given considerable insight into the host resistance mechanism [6]. Immune related genes have been identified as being determinant in a complex interplay of numerous genes [6]. Based on survival times of laboratory rabbits inoculated with different virus isolates Fenner and colleagues defined five virulence grades 1–5 [4]. Virus isolates of virulence grades 1–2 have over 95% case fatality rates with survival times of less than 13 days for grade 1 or up to 16 days for grade 2 virus. Rabbits infected with grade 5 virus survive while grades 3 and 4 average survival time is 17–29 and 29–50 days respectively. Experimental studies in the decades following virus release showed grade 3 viruses to be most common in both Australia and Europe. Changes in virus virulence occurred rapidly, with less virulent strains detected after one year [4] and a comparison of results from Australia and Europe showing a common trend and highlighted the importance of insect vectors and weather conditions on selection of virus attenuation [4, 5, 7, 8].

#### *2.1.2 Control and prevention*

Myxoma virus (MYXV) is the virus of reference for the *Leporipoxvirus* genus. Poxvirus virions are large with a brick or ovoid shaped morphology (ViralZone). Replication occurs in the cytoplasm of infected cells. (reviewed in Fields Virology).

The MYXV genome is a linear double stranded DNA molecule of 161.8 kb encoding 171 open reading frames (ORFs). The ends of genome are covalently closed forming hairpin terminal loops. The genome of MYXV Lausanne was sequenced completely in 1999 by Cameron et al. [9] and contains inverted terminal repeats (ITRs) of 11.5 kb encoding 12 genes which are therefore diploid. The MYXV genome contains an abundance of genes with potential immunomodulatory functions contained largely in the ITRs while genes the central core of the genome are required for virus replication, transcription and morphogenesis (**Figure 1**). Genes found in this central region are highly conserved among the poxviruses [12].

There is no treatment for the disease therefore only preventative measures are effective. As the disease is spread most commonly by biting insects (fleas, lice, ticks and mosquitos), insect control has fundamental importance, being the first line of defence on the farm setting. The introduction of new individuals on to a farm or wildlife area should be preceded by quarantine measures and animals showing clinical signs should be isolated. However, the only effective measure for virus control is currently vaccination.

Several vaccines exist against myxomatosis and these can be divided into two types; heterologous and homologous with both types being live attenuated vaccines. The heterologous vaccines use RFV (also denominated Shope fibroma virus). RFV can be used to provide protection against myxomatosis [13] as it is immunologically related to MYXV [14]. However, protection lasts a short time and the heterologous vaccines are considered less immunogenic. Heterologous vaccination is often used to vaccinate juveniles for the first time whilst subsequent vaccinations are administered using homologous vaccines. The homologous vaccines are based on laboratory MYXV strains that have been passaged in embryonated eggs or cell cultures until an attenuated phenotype is attained [15–17]. Several homologous vaccine strains are available in Europe (eg. SG33, MAV, Borgi, Leon162 among others). Protection with homologous vaccines offers longer lasting immunity than with the heterologous vaccines. Myxoma virus causes immunosuppression in rabbits [18] and one of the main drawbacks to the use of homologous vaccines may be associated with such immunosuppression [18, 19] that could exacerbate underlying subclinical infections.

The genome sequences of various vaccine strains have been studied. While there is no consensus on individual gene mutations that cause the attenuated phenotype in vaccines, several have large regions of the genome absent when compared to the parental strains (**Figure 1**) [11, 20, 21]. For example, the MYXV vaccine strain SG33

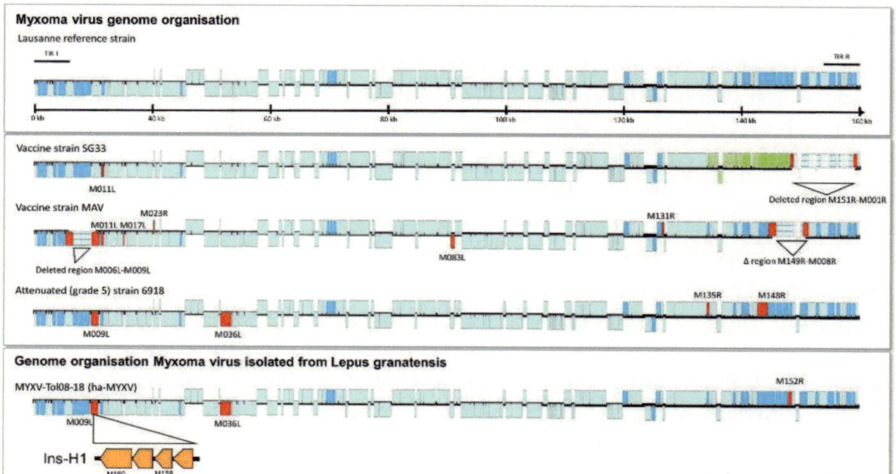


**Figure 1.**
*Schematic representation of MXYV genome organization. Genes below the black line are designated L and those above the line are R. Immunomodulatory proteins of MYXV shown in darker shade of blue [9, 10]. Genes colored green indicate homology with California (MSW/ MSD) strains [11]. Genes showing major disruptions in attenuated strains shown in red and deleted regions indicated. Ins-H1 insertion in MYXV-Tol08-18 shown in orange.*

has a genome 13.5 kbp shorter than Lausanne. The deleted region contained genes encoding for M150-M001 proteins (13 genes) including several genes expressing immunomodulatory proteins. The region affected contains a large part of the ITR and the right end of the central genomic region, while other mutations are also present in the genome notably in the open reading frames M011 and M077. SG33 was derived from a French field isolate and subsequent analysis of the genome revealed it to be a recombinant consisting of genomic sequences of the south American Lausanne and the Californian MSW/MSD strains [11]. The MAV vaccine strain (derived by serial passage of strain MSD isolated during an outbreak in California [22] has a 14.2 kbp deletion with respect to Lausanne, affecting both terminal regions flanking the ITRs (genes M006L-M009L and M148R-M008L/R). Interestingly, one European field strain has been sequenced that also contains a large deletion including part of the ITR, isolate Munich-1, this strain is described as virulent. Therefore, the precise mechanisms of attenuation of vaccine strains remain undefined however, the role of the large deletions that include genes involved in immunoregulation seems evident.

The publication of the MYXV genome coupled with advances in recombinant DNA technology and *in vitro* poxvirus manipulations paved the way for the design of recombinant attenuated strains and the development of potentially safer vaccines and may offer the possibility of reducing immunosuppressive effects. The deletion of virulence factors and the development of viruses with attenuated phenotypes [23, 24] has provided key information for the possible development of recombinant vaccines based on targeted attenuation.

Novel recombinant bivalent vaccines have been developed using MYXV as a vaccine vector that enable simultaneous vaccination against MYXV and other pathogens in domestic [25, 26] and potentially wild rabbits [27, 28]. The Spanish field strain 6918 showed great promise as a potential vaccine candidate in wild rabbits. A field trail of this vaccine showed that 50% of contact rabbits generated antibodies over a 32-day period. This isolate has also been used as a vector to express the capsid protein of rabbit haemorrhagic disease virus. The potential use of MYXV as a vaccine vector is not limited to use in rabbits. Indeed vectors have been tested in different species [29–31] including humans for use as an oncolytic virus vector (reviewed in [32]).

Outbreaks of myxomatosis still occur in farmed rabbits although efficient vaccines have been available since the 1980s. One major problem is that field strains are endemic in feral rabbit populations. There is therefore a constant source for virus introduction onto farms through uncontrolled insect vectors. Such outbreaks lead to doubts over the efficacy of current vaccines and the causes of these vaccine breaks have been investigated. Sequence data of field strains causing farm outbreaks show that the virus has not changed sufficiently to render vaccines outdated, indeed under laboratory conditions a traditional vaccine strain proved protective if the vaccinated rabbit generated antibodies [33]. However, the route of administration appeared to be a crucial factor. In this study a group of animals vaccinated subcutaneously failed to seroconvert and were susceptible to fatal infection, while animals vaccinated intradermically showed seroconversion and were protected. The key to controlling such phenomena on the farm setting is surveillance of antibody response following vaccination.

The two strains of the virus initially released in Europe and Australia have been completely sequenced. The reference strain (isolated in Brazil in 1949) known as Lausanne was released in Europe, while the Standard Laboratory Strain (SLS) also termed Moses strain was isolated in Brazil in 1910 and maintained by passage prior to its release in Australia (Moses 1911, reviewed in [5]). The fact that the date, locations, and viral sequence of released strains are known allows for the comparative study of past and present MYXV sequences. Such precise data is unique and have allowed for continental scale virus evolution studies [21, 34, 35]. These studies have demonstrated that MYXV showed high mutation rates, frequent loss of ORFs due to nucleotide insertions or and deletions [21, 34, 35]. The evolution of complete genome sequences of MYXV strains over more than 70 years coupled with the data on virulence and virus phenotypes provided by disease monitoring programs [4] provide a powerful tool for the detailed analysis of the molecular mechanisms of virulence and attenuation. In addition, the study of recombinant knockout viruses, viruses with individual genes removed under laboratory conditions, and subsequent analysis of changes in virulence associated with gene knockouts allows the determination of these mechanisms with much greater precision (reviewed in [21]). Understanding the molecular mechanisms of attenuation will be crucial for the design of better control measures of MYXV.

#### *2.1.3 Molecular mechanism of attenuation*

Phenotypic studies of MYXV isolates demonstrated the emergence of attenuated field strains. Analysis of sequences of attenuated strains has revealed that virus attenuation is multifactorial. Such studies reveal there are varied pathways to the evolution of attenuation [2, 35]. Several genes have been shown to be involved but no single common mutation or group of mutations account for similar virus phenotypes. Genome wide analysis of MXYV strains isolated between 1952 and 1999 demonstrated the broad range of genes involved in MYXV evolution [34]. Strains of known virulence grades were included in such studies and although viruses shared virulence grades no common mutations between them were observed [34, 36] with similar results being obtained for Australian and European isolates [35, 36]. A subsequent study of isolates obtained between 2000 and 2021, (following the emergence of RHDV) showed a drastic change in virus evolution, echoing the effects that RHDV had on rabbit populations and highlighting the adaptability of MYXV to persist in the face of ever changing circumstances [37]. A more pronounced virus evolution was detected but once again the complexity behind attenuation/virulence were noted [37] as were the effects of RHDV and climatic conditions on the evolution of spread and transmission [37, 38].

The complexity behind the molecular explanation of attenuation is further exacerbated with the knowledge that individual or combination gene knockout studies have not precisely replicated attenuation in recombinant strains [39]. For example, of the Australian attenuated isolates, Uriarra contains an indel disrupting the M005L/R gene, while Meby codes for a truncated M153R gene. Both genes products play roles in immunomodulation [40, 41] and recombinant viruses with these genes deleted are less virulent (M005 highly attenuated, M153 moderately, [40, 42] respectively). However, reversion studies (replacing defective genes with corrected homologs in the genetic background of the parental attenuated strain) failed to show differences in the virus phenotype. The role of additional subtle mutations occurring in other parts of the virus genome although more difficult to decipher may hold the key to understanding attenuation completely. In the case of European isolates, attenuated strains have also been identified, e.g. Nottingham4-55/1 and the Spanish isolate 6918. Nottingham4-55/1 contains a truncated M150R gene. M150R (also termed myxoma nuclear factor (MNF)) is an immunoregulatory/host range protein [43] and one of the poxviral ankyrin-repeat (ANK-R) protein superfamily members. Recombinant virus with an MNF deletion is attenuated [44]. In addition to being attenuated, the Spanish isolate 6918 was poorly transmitted to contact rabbits demonstrating an additional phenotypic characteristic making it a possible safe vaccine candidate for use in wild rabbits [27, 28, 45]. Genome sequencing of isolate

6918 revealed mutations in four genes (M009L, M036L, M135R and M148R) leading to potential truncation of expressed proteins [46] (**Figure 1**). The M135R and M148R genes are likely candidates to affect virulence, while mutations in M009L and M036L genes have also been detected in virulent MYXV strains [47] however, the precise relation of these genes with attenuation has yet to be demonstrated. To date reversion studies using European strains have not been documented.

#### *2.1.4 Diagnosis*

In Europe the disease had a profound effect on rabbit populations including extinctions at local levels [48, 49] and much wider ranging ecological effects [49, 50] especially in the Mediterranean basin where the rabbit is considered a keystone species [51, 52]. Surviving rabbits show immunity [53] therefore disease maintenance requires sufficient susceptible naive juvenile rabbits and outbreaks therefore occur in temporal peaks which are also dependent on the availability of insect vectors which help to spread the virus (reviewed in [5]). In addition to the ecological effect, the rabbit industry (meat and fur) suffers substantial economic losses each year due to rabbit deaths and costs of control measures [54, 55].

Initial diagnosis may be based on the characteristic symptoms presented, however, disease caused by attenuated or respiratory strains maybe mistaken for bacterial infections. Laboratory diagnosis is therefore important and several methods are available for this task (reviewed in [56, 57]). Isolation of the virus in susceptible cell cultures (e.g. RK13 rabbit kidney cells) gives characteristic cytopathic effect but requires several days for confirmation and is therefore accompanied by serological or molecular techniques such as the detection of virus genome by PCR [20]. The most common samples for virus detection are tissue (eyelid, skin lesions, lung etc), swabs (conjunctival or genital) and sera for antibody detection. These techniques maybe complemented by electron microscopy, histology, agar gel immunodiffusion, fluorescent antibody test among others [57]. Detection of the virus genome requires extraction of viral DNA and subsequent PCR analysis and confirmation by sequencing may also be used. Targeted regions vary and include M153R [58] and M135R [59] or M005L/R gene within the TIRs (and therefore diploid) for qPCR [60] or conserved genes such as M022L [61] or M071L [20, 57] for conventional PCR analysis. Such analyses serve to confirm the presence of MYXV DNA, however, additional analysis of more regions are required to distinguish wildtype from vaccine strains [20] or in the absence of genome sequencing, to generate information regarding phylogeny [62, 63]. TIR regions in poxvirus genomes have been shown to be more variable [64], and RFLP analysis of long range PCR-amplified TIR regions from MYXV positive samples also allows for the molecular differentiation of strains [65].

Serological tests exist for the detection of anti-myxoma virus antibodies [19, 66, 67]. This analysis is of particular importance in farmed rabbits where vaccination is used.

#### *2.1.5 MYXV jumps species to the Iberian hare (Lepus granatensis)*

Myxoma virus is traditionally termed as being specific to the European rabbit. Soon after its release in Europe however, isolated cases were described in hare species (reviewed in [4]). More recently myxomatosis has been described in hares in the UK [68]. Experimental infections of hares (*Lepus europaeus*) with MYXV failed to result in disease [4] and natural cases seem to have been sporadic as no large-scale infections were described. These results do, however, demonstrate the susceptibility of hare species to MYXV infection, at least in certain circumstances. In spring 2018 the situation changed dramatically with reports of widespread infections in the Iberian hare (*Lepus granitensis*) occurring on the Iberian Peninsula [69–71]. Over 300 cases were reported and confirmed as positive for MYXV thus being the first large scale infection of hares with MYXV. The outbreak continued and spread over the Iberian Peninsula throughout 2018–2020 [72]. The impact this virus will have on the Iberian hare population has yet to be seen and must be continually monitored so that suitable control measures can be adopted. The virus has also been detected in wild rabbits (*Oryctolagus cuniculus algirus*) and commercial rabbits highlighting the need for control and monitoring [73].

Genomic analysis of the MYXV infecting the Iberian hare population showed the genome to contain an additional 2,8 kbp with regards to the reference strain Lausanne (**Figure 1**). How MYXV gained this genomic region is unknown, but homologous recombination has been demonstrated as a frequently occurring mechanism for poxviruses to gain selective advantage through the acquisition of genetic material from coinfecting viruses [74–76]. The genomic region present in the hare- infecting MYXV contained 4 full open reading frames which showed homology to MYXV genes M060R, M061R, M064R and M065R. The homologous region within the MYXV genome contains 6 genes, M060R-M065R, therefore homologs to M062 and M063 are missing from the hare specific genomic region. The nature of the ORFs included in the insertion led to two possible explanations: a duplication event of genes M060-M065 had occurred with the subsequent loss of ORFs M062 and M063 [70], or the capture by homologous recombination of this region from an as yet unidentified poxvirus [70, 71]. The exact cause of the species jump has yet to be explained, although the determination of the function of the gene products from within this region is sure to shed light on this phenomenon.

## 3. *Lagovirus* (family *Caliciviridae*)

There are two devastating diseases caused by lagoviruses that effect lagomorphs. Rabbit haemorrhagic disease (RHD) and European Brown Hare Syndrome (EBHS), as the names suggest, both were considered species specific. However, recent findings require that this view be revised. Both diseases have been endemic in Europe since the first descriptions in the 1980s (reviewed in [77, 78]). Due to the economic and ecological importance of the European rabbit, RHD has been the subject of most research. But the recent emergence of a lagovirus with broader host range, associated ecological concerns and advances in the molecular biological tools available for the study of these diseases, is changing this dynamic.

### 3.1 Rabbit haemorrhagic disease virus (RHDV)

#### *3.1.1 RHDV GI.1*

Rabbit haemorrhagic disease first came to light in China in 1984 in angora rabbits imported from Europe in what appeared to be the emergence of a novel highly virulent disease that rapidly killed thousands of domestic rabbits. The disease spread throughout China and Korea and reached Europe (Italy) in 1986 (reviewed in [77, 79]).

The first report of cases in Spain, the native home of the European rabbit was in 1988 [80] in domestic rabbits and the disease soon caused epizootic episodes in wild rabbits [81]. The reduction of the wild rabbit population was severe [82] and this had direct effects on specialist predator species such as the endangered Iberian lynx (*Lynx pardinus*) or Spanish Imperial Eagle (*A. adalberti*) [52, 83]. Such were the effects of the disease on rabbit populations in Europe that tests were carried out for the use of the disease as an additional biocontrol measure against the rabbit pest in Australia, which was recovering following the initial success of myxomatosis in control efforts [84].

RHDV, the etiological agent of RHD, remained endemic in Europe for more than 30 years with a single serotype (RHDV GI.1) prevalent until 2010. Rabbits that survive infection generate a strong long lasting immunity with detectable anti-RHDV antibodies in sera. Young rabbits are not susceptible to the disease caused by RHDV GI.1, however, they may be infected and the mechanism of resistance is not fully understood. The economic effects of the disease in the rabbit farming sector are considerable. Inactivated vaccines (e.g. RHDV infected liver homogenates treated with β-propiolactone) were developed [85, 86] and proved successful in stemming mortalities in commercial rabbitries although due to the natural reservoir of RHDV in wild rabbits, control measures must be consistently maintained. With regards to control it is important to indicate that direct contact between rabbits [87] and fomites (contaminated food and bedding) play an important role in farm outbreaks. In natural infections, the faecal-oral route is considered the preferential mode of virus transmission [88] reviewed in [78]). In the wild, rabbit carcasses are also a major virus source with spread being facilitated by predators and carrion feeding insects [89, 90].

### *3.1.2 RHDV GI.2*

In 2010 atypical outbreaks of RHD were detected. The virus responsible for these outbreaks was initially recognized in France [91] and termed a new variant of RHDV, where outbreaks affecting vaccinated rabbits caused concern. The variant was later detected in Spain, where isolates of the virus were used to show susceptibility of young rabbits, demonstrate major antigenic differences with regards to RHDV GI.1 and the presence of virus in the intestine [92]. Another considerable difference with the disease caused by this virus when compared to classic RHDV GI.1 is the level of mortality. RHDV GI.1 typically shows mortality rates of 80–90% in adult rabbits [93] with no mortality in kits, while the variant RHDV showed approximately 10% in adult rabbits and up to 50% in kits. The terms variant and RHDVb were first used to identify this virus [91, 92]. Subsequent publications used the terms RHDVb and RHDV2. In order to avoid confusion and bring order to the nomenclature system for the lagoviruses, Le Pendu et al., put forward a new scheme where by RHDVb/RHDV2 was termed *Lagovirus europaeus/GI.2/* (GI.2), and RHDV was termed *Lagovirus europaeus/GI.1/*, (GI.1). The proposed nomenclature and classification system allows systematic definition using phylogeny and genetic distances to define isolates, includes pathogenic and non-pathogenic lagoviruses and allows for the incorporation of as yet unidentified virus sequences [94].

Given the novel characteristics of RHDV GI.2 it was unsurprising that the virus continued to spread. By 2013, the virus had been detected throughout France, Spain, Portugal and was present in Italy [95–97]. RHDV GI.2 and has since spread to pandemic proportions and has been detected on the continents of Europe, Africa, Asia, Australia and North America.

The lack of full cross protection induced by previous contact with RHDV GI.1 strains contributed to the rapid spread of RHDVGI.2 in Europe [98, 99], resulting in high mortality rates among wild populations soon after its emergence. Therefore, vaccines based on RHDVGI.2 containing inactivated liver extracts have been produced to aid in control of the disease in domestic rabbits.

The observation of partial protection between GI.1 and GI.2 was supported by data from RHDV GI.2 infections in Australia where mortality was detected in

RHDV domestic vaccinated animals [100] and contributed to its rapid spread in the wild [101]. Currently, it appears as though RHDV GI.2 has become the predominant RHDV on the Iberian Peninsula [102, 103], and also on mainland Australia replacing endemic strains of RHDV GI.1 [100]. While this may lead to environmental and economic benefits in Australia [104] the establishment of this virus in Europe poses an important problem for the conservation of the European rabbit, particularly in smaller populations [105, 106] and for the preservation of reliant predators [107].

### 3.2 Rabbit calicivirus (RCV) GI.3 and GI.4

The presence of anti-RHDV antibodies in rabbit populations is indicative of circulation of RHDV. However, in 1995, antibodies reactive with RHDV were detected in rabbits that showed no clinical signs of RHDV infection [108]. This finding was substantiated by the identification of a non-pathogenic calicivirus (termed rabbit calicivirus - RCV) [109]. RCVs were subsequently detected in Australia [110] and France [111]. RCV lagoviruses, are genetically related to RHDV although they demonstrate different cell tropism and are apathogenic. While RHDV target organs are lung, liver and spleen, RCV was found predominantly in the intestine. The presence of RCV has been argued as having a protective effect on rabbit populations facing RHDV outbreaks and may have been a determining factor in the speed of spread of RHDV in Europe and Australia [112], however the levels of protection vary and are likely dependent on differing levels of cross reactive and timing of infections [111, 113].

### 3.3 European brown hare syndrome virus (EBHSV) GII.1 and Hare calicivirus (HaCV) GII.2

European Brown Hare Syndrome Virus (EBHSV) GII.1 has been detected in many European countries and Argentina having emerged in Sweden in the early 1980s and Denmark in 1982 (reviewed in [114–121], although retrospective studies have demonstrated that the virus was present before this time [122, 123]. EBHSV causes disease in brown hares (*Lepus europaeus*) and mountain hares (*Lepus timidus*) [114, 124], while the eastern cottontail (*Sylvilagus floridanus*) is also susceptible [125] but not the European rabbit (*Orytolagus cuniculus*) [126]. The disease symptoms are similar to those observed for RHDV, most notably the disease is acute and may show respiratory and nervous symptoms and epistaxis [127], the liver shows severe necrotizing hepatitis although differences between the diseases have also been reported [128]. Similar to RHDV GI.1, EBHSV does not infect young hares [121, 124].

The identification of hare caliciviruses (HaCV) GII.2 related to EBHSV have further shown the large diversity and complexity that exists for this genera of virus [129–132]. HaCV has been detected in duodenum or faeces of healthy hares in Italy, France, Austria, Germany and Australia (also weakly detected in liver) [129–133]. The virus is considered to be non-pathogenic based the health status of animals from which the samples were obtained and the target organ, however, virulence phenotypes have not been published. Expanding our knowledge on these viruses will undoubtedly help decipher their role in protecting populations of lagomorphs from pathogenic viruses and in the emergence of novel viruses through recombination events.

As lagoviruses EBHSV and HaCV have the same genome organisations (**Figure 4**) and virion morphologies as RHDV, however they form separate genetic groups following phylogenetic analysis and are antigenically different from RHDV [123, 134–137]. RHDV GI.2 has also been detected as highly virulent in different hare species [138] either as a result of spill-overs from closest rabbit populations or from hare to hare transmission. These findings have highlighted the need for studies into virus cross species infections in lagomorphs [130, 132]. Recent detection of RHDV GI.2 and EBHSV GII.1 recombinant viruses [139] highlight the capacity for evolution of the lagoviruses and indicate the importance of continued vigilance in order to protect vulnerable lagomorph species.

### 3.4 Virion, genome organization

Lagoviruses are non-enveloped icosahedral single-stranded positive-sense RNA viruses [87, 140, 141]. As well as the morphological features of virions all lagoviruses share a common genome organisation, have conserved genomic features and express two ORFs as shown in **Figure 2**.

ORF 1 expresses a polyprotein that is processed to form non-structural mature peptides and VP60 the major structural capsid protein (also termed VP1 in the literature). ORF 2 encodes VP10 a minor structural protein, also termed VP2. Lagoviruses also share conserved polyprotein processing sites (**Figure 3**) and. Transcribe a VPg-linked subgenomic RNA from which VP60 and VP10 are expressed. The ORFs for VP60 and VP10 overlap and expression of VP10 occurs via a termination/reinitiation mechanism (**Figure 3**).

RHDV GI.1 is the lagovirus that has been most extensively characterised and is therefore the virus of reference for this genus. The virus genome is approximately 7.4 kb in length. Viral particles are small (35–40 nm diameter) and contain genomic (gRNA) and subgenomic RNA (sgRNA) which is collinear with the 3′ end of the genomic RNA [143–145]. Both RNAs are polyadenylated at the 3′ end. At their 5′ region they are covalently linked to the VPg (virus genome-linked) protein [146] (**Figure 2**) which may act as a substitute for cap during RHDV translation [147]. Genomic RNA contains two open reading frames (ORFs): ORF1 codes for a polyprotein of 257 kDa, which after a post-translational cleavage by a viral protease results in 7 non-structural proteins (p16, p23, p37 (NTPase), p29, p13 (VPg), p15 (protease),

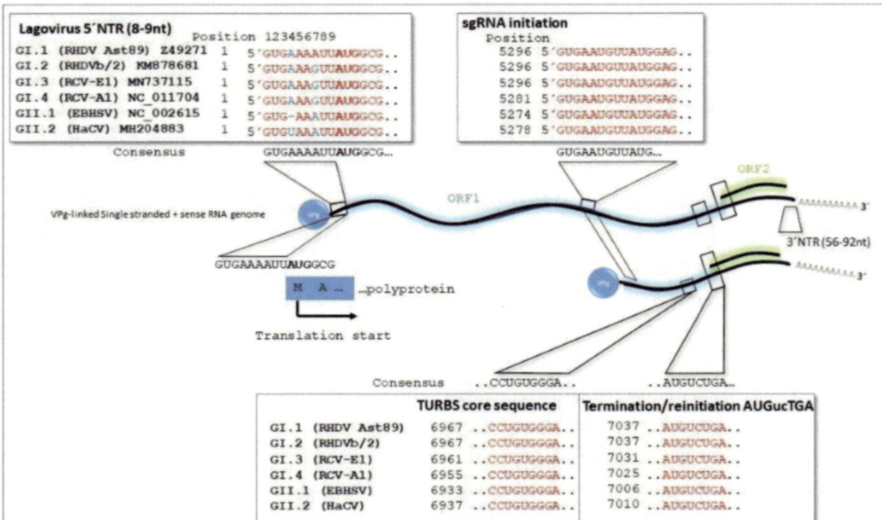


**Figure 2.**
*Schematic representation of a lagovirus genome and the main genomic features shared amoung the lagoviruses. Fragments of sequence alignments from indicated isolates show: 5′ nontranslated region (NTR), polyprotein translation start site, subgenomic RNA initiation sites, termination upstream ribosomal binding site (TURBS) core sequence and termination/reinitiation site required for VP10 translation [142].*

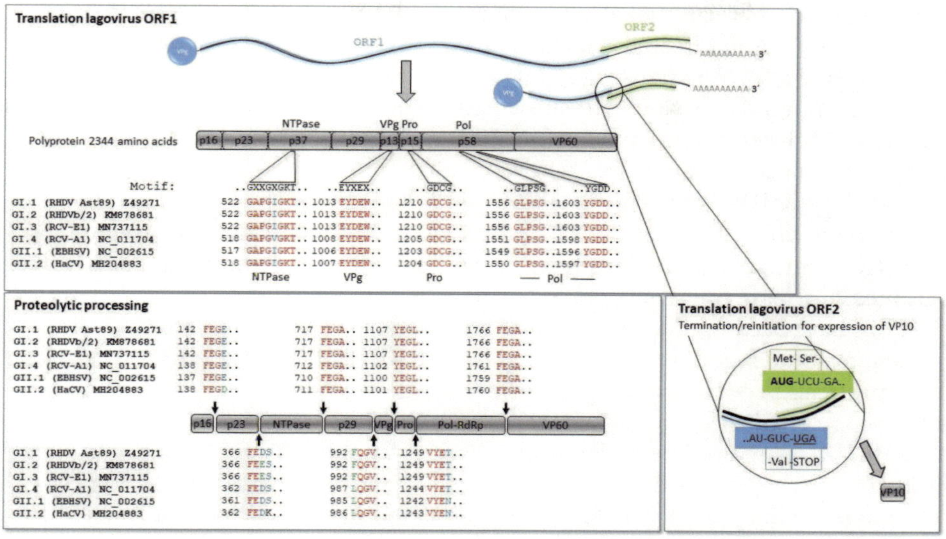


**Figure 3.**
*Schematic representation of conserved motifs located in the ORF 1 encoded polyprotein and amino acid alignments showing conserved proteolytic processing sites among lagoviruses.*

p58 (RNA-dependent RNA polymerase) (also termed non-structural proteins (NS) 1–7) and the major capsid protein (VP60) [144, 145, 148]. The RNA sequences encoding ORF1 and ORF2 overlap by 20 nucleotides. The reading frame of ORF2 is shifted with respect to ORF1 and codes for the minor capsid protein (VP10) which is 10–12 kDa. VP10 (also termed VP2 in some publications in accordance with norovirus nomenclature) is translated by a termination/reinitiation mechanism [142], dependent on a TURBS (termination upstream ribosome-binding site) element (core sequence GUGGGA) located within the VP60 coding sequence [149]. The role of VP10 protein has been linked with viral replication regulation and promotion of apoptosis for the liberation of virions from infected cells [145, 150]. The biological role of non-structural proteins has been studied by sequence analysis and functional studies, as well as being based on previous knowledge gathered from members of the *Picornaviridae* family [143, 145, 151]. Four of the nonstructural proteins have well defined functions, namely the RNA helicase (p37) [151], the virus genome linked protein VPg (p13), 3C-like protease (p15 or Pro) and the RNA-dependent RNA polymerase (RdRp) [146, 152, 153] (**Figure 3**). RdRp has been shown to catalyze VPg uridylation [146, 154]. The RHDV protease (Pro) [155] catalytic site has been mapped by site directed mutagenesis [155] and the target viral peptides have been identified [156]. The RdRp functions in replicating the viral RNA and VPg uridylation [154]. While the precursor (Pro-Pol) shows activity, the processed mature form shows increased capacity in polymerase function [154]. Expression of recombinant RHDV RdRp in transfected RK13 cells leads to Golgi membrane reorganization [157]. The crystal structure of this protein has been elucidated [158] allowing insights into the structural-functional mechanism involved in virus replication. The intracellular location of the RHDV non-structural proteins p16, p23 and p29 have been recently determined [157], however, their functions are still unknown.

### 3.5 VP60 major structural virion component

By far the most well characterized of the RHDV proteins is VP60. It is the major structural component of virions, responsible for receptor binding, is highly immunogenic and the target of neutralizing antibodies. VP60 is expressed from subgenomic RNA in infected cells facilitating sufficient levels for virion assembly. High levels of sgRNA make it a good target for diagnostic RT-PCRs and its variability due to immune selection befit subsequent sequence analysis. Sequence data from partial and full-length fragments of VP60 encoding RNA have been the basis for comparative sequence analysis.

Although RHDV is not cultivatable *in vitro* the expression of VP60 and the formation of virus-like particles (VLPs) makes recombinant vaccine production an attractive alternative and ethically acceptable substitute for traditional inactivated vaccines which are prepared from infected liver homogenates.

VLPs have been used to characterize this protein structurally, determine its antigenicity and study receptor binding [159–164]. Additionally, purified wildtype virus particles have been used to determine an atomic model by cryo-electron microscopy [165].

RHDV GI.2 isolates are antigenically distinct from RHDV GI.1 and agglutinate human erythrocytes, to differing degrees, [92, 166–168]. Hemagglutination is dependent on the presence of ABH blood group antigens and these molecules have been proposed as receptor components of the binding pathway [169]. Indeed, the species specific nature of these molecules may explain susceptibility to different lagoviruses [170].

Virions are composed of 90 dimers of VP60 which form 32 cup-shaped depressions on the surface of virions (from which comes the name Calici- from the latin calix or cup) [165, 171]. Each monomer of VP60 consists of a shell (S) domain, which is buried inside the virion structure, and flanked by the N-terminal arm (NTA) and the so-called hinge which links the S domain to the protruding (P) domain that correspond to C-terminal region and is exposed on the virion surface [166, 172–174]. The hinge allows P and S domains a certain grade of flexibility on their tridimensional disposition [165]. The P domain contains determinants for virus-host receptor interactions and antigenic diversity [165, 166]. In addition, this region can be further subdivided into the subdomains P1 and P2, P2 contains the hypervariable region of the protein which allows for the selection of variants that can escape immune detection or antibody neutralization.

The marked antigenic differences observed between RHDV GI.1 and RHDV GI.2 [168], explain the lack of efficient protection against RHDV GI.2 afforded by RHDV GI.1 inactivated vaccines [91, 92], as well as, the capacity of RHDV GI.2 to overcome immunity derived from natural infections with RHDV GI.1 [101]. Since the emergence of RHDV GI.2 the development of diagnostic tools has been a very important issue and different techniques and methodologies have been improved for specific detection of RHDV GI.2 [168, 175–177].

### 3.6 Evolution: genotypes, antigenic variants

Although perceived as a novel virus (GI.1) in the initial 1984 outbreak, retrospective serological analysis [108] and sequence analysis [178–181] have cast doubt on this assumption. Molecular clock analysis suggest the virus emerged before it was officially detected [182]. These studies back the theory that RHDV or similar viruses existed in less virulent forms or went unnoticed and were circulating in the rabbit population long before the emergence of RHDV as a major concern [183]. Whether GI.1 and GI.2 arose from non-pathogenic viruses, through recombination events, or through species jumps remains to be determined [184–186].

Genetic diversity among pathogenic RHDV isolates has been the subject of intense study with the capsid protein (partial or complete) being the main target. Recent advances in sequencing techniques has allowed full genome analysis in retrospective studies and demonstrate the importance of analysing complete genomes [187, 188].

Studies on the evolution of RHDV GI.1 in Europe showed low levels of sequence variation [183, 189] in the years following its emergence although it was possible to define distinct phylogenetic groups. In the 12 years following the emergence of RHDV GI.1 in France six genogroups were identified [190, 191] (now termed GI.1a-d). Genogroup 6 or RHDVa (GI.1a) is an antigenic variant first isolated in Italy [192]. Although antigenically distinct with a sequence similarity in the VP60 gene of 93%, GI.1 based vaccines provided protection against this highly pathogenic virus [191, 192].

The evolution of RHDV has followed a different trend in Spain compared to the rest of Europe. Only GI.1d strains (previously genogroup 1) have been detected in samples collected between 1988 and 2010. This may be due to a smaller number of samples being analysed, however, unlike in the rest of Europe, GI.1 strains were still being detected in the Iberian Peninsula in 2010. RHDVa was detected once in Spain in domestic rabbits [193] but does not appear to have been widespread and has not been detected in wild rabbits. In Australia the Czech strain V-351 (an RHDV GI.1) escaped from testing facilities in 1995. The sequence of virus circulating in Australia changed little in the first years following its escape [194], however, the study of RHDV sequences spanning the subsequent 16 years revealed its evolutionary tendency on this continent. Although subject of more than 3500 independent releases during this time, positive selection suggested the evolution of strains capable outcompeting freshly released strains and spreading in the presence of non-pathogenic rabbit calicivirus (RCV)-A1 (also GI.4a) [195]. In order to more effectively compete against the presence of RCV-A1, the antigenic variant RHDVa-K5 (GI.1a-K5) was released in 2014. Australian RHDV GI.1 strains gained virulence reflecting the selection of viruses [196] based on effective transmission, the immunological status of existing population, landscape and weather amongst others.

Phylogenetic analysis suggested that RHDV GI.2 is genetically distant from RHDV GI.1 and is more closely related to RCV apathogenic viruses [91, 92]. The origin of RHDV GI.2 is unclear and cannot be explained by genetic evolution from previously described lagoviruses or recombination of existing strains [197, 198]. Silverio and colleagues [197] using sequences analysis and mean substitution rates have estimated the presence of a common ancestor for GI.2 just a few years prior to its first detection [197]. Experimental infections have demonstrated that RHDV GI.2 has gained virulence showing higher mortality rates in both adults and kittens [98, 199], than the strains obtained in 2011 or 2012 soon after its emergence [98, 199]. This could indicate an evolutionary tendency of the virus, indeed, Capucci and colleagues (2017) [199] hypothesized that this could be similar to what occurred for RHDV GI.1 strains in Australia [196], RHDV GI.2 having evolved in their natural hosts and, since their emergence in 2010, selection pressure may have favoured strains with higher pathogenicity.

### 3.7 Recombination and lagoviruses

Recombination, along with mutation, is an important mechanism for the evolution of RNA viruses since it uses existing genetic diversity to create new genomic combinations. Novel RHDV virus genomes arising from natural recombination of existing strains have been detected in RHDV isolates [200–204]. Recent genomic analysis reveals that these events are more common than once envisaged [200]. and due to the high frequency of occurrence [195, 197, 200–203, 205] this might indicate an important role in their origin and emergence of pathogenicity. Indeed, Silverio and colleagues [197] indicated that the occurrence of recombination fits both theories currently proposed for the emergence of pathogenicity in lagoviruses: (1) the evolution from a pre-existing non-pathogenic virus that acquired pathogenicity or (2) following a species jump [198]. Especially in the case of the highly frequent recombination events observed for the novel RHDVGI.2 and its ability to cross the species boundaries [99, 138, 206–208]. This also has been supported by Lopes and colleagues [187], who related the detection of a recombinant RHDVGI.2 in an Iberian hare with a species jump [187].

The number of combinations of genomes being observed is striking and reveals a complicated panorama. How these events shape the spread and pathogenicity of RHDV remains to be determined.

In recent years, recombination events in different regions of RHDV genomes have been identified. The existence of recombinant RHDV that contain structural genes from RHDV GI.1 and non-structural genes belonging to non-pathogenic lagovirus (GI.4) have been identified [202]. In some cases, the recombination event occurred inside the non-structural region [187]. Two types of RHDV GI.2 recombinant strains have been identified, both with their structural proteins VP60 and VP10 originating from GI.2, while their non-structural proteins originated from GI.1 or non-pathogenic strains (GI.4) [103, 187, 197].

Recombination has been reported in Iberian GI.2 genomes, with a breakpoint at the RdRp/VP60 boundary (**Figure 4**) within ORF1. This breakpoint was associated with several independent recombination events involving non-pathogenic strains, GI.1 and/or GI.2 resulting in different genomic combinations that persisted in the Portuguese wild rabbit populations [187, 197], and recombinant strains detected in Azores Islands and Australia [103, 202, 209]. The homology between the subgenomic RNA (encodes structural genes) and the 3′ end of the genomic RNA (encodes non-structural and structural genes) generates a hotspot of recombination at the junction between non-structural and structural genes [200, 202]. Silverio and colleagues [197] identified new RHDVGI.2 recombinants with a recombination breakpoint located near the p16-p23 (NS1/NS2) boundary (nucleotide positions 355–471), [197]. Also, they detected the occurrence of triple recombinants constituted by the NS1 non-structural protein similar to a nonpathogenic lagovirus, a

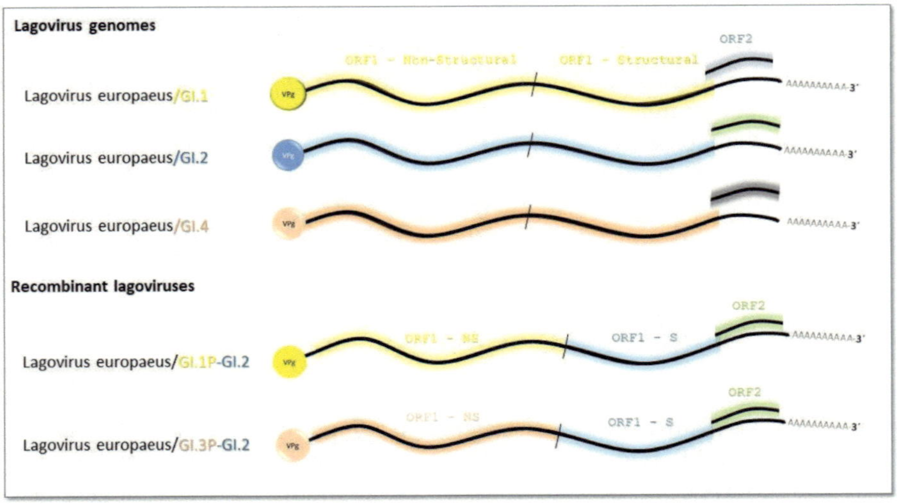


**Figure 4.**
*Schematic diagram of recombinant lagovirus genomes. In the example shown recombinant Lagovirus europaeus/GI.1P-GI.2 indicates a recombination detected between a GI.1 and a G2 strain; P standing for polymerase.*

GI.1b backbone for the remaining non-structural proteins and GI.2 as the donor for the structural protein. Mutations in NS1 sequences has been implicated as a factor in increased virulence of GI.1 isolates [196].

### 3.8 Cross species infections of lagoviruses in lagomorphs

Prior to the emergence of RHDV GI.2, lagoviruses were considered species specific, RHDV was specific to the European rabbit and EBHSV was specific to the European brown hare [126]. However, increasing numbers of RHDVGI.2 infections in different species of rabbits and hares have been reported casting doubt on this clear differentiation. Initially considered as transient spill-over events, more widespread infections of GI.2 in a number of hare species have raised concerns that this virus has an ampler species tropism. The first detection occurred in 7 Cape hares (*Lepus capensis* var. mediterraneus) with lesions consistent with those observed in RHD infections [99]. This study demonstrated for the first time that a *Lepus* species showed susceptibility to RHDV GI.2 and a clear difference with regards to susceptibility to RHDV GI.1 in this species given that RHDV GI.1 had previously been endemic in rabbits in the same geographical area for 20 years [99]. Subsequently RHDV GI.2 was determined as the cause of death of a captive Italian hare (*Lepus corsicanus*). In this species virulence was more limited, only infecting a single hare on the compound [206]. Velarde et al. [138] described the first infections of RHDV GI.2 in wild European brown hares (*Lepus europaeus*) detected in Italy (n=1) and Spain (n = 2). Serological data from wild hares presented in this study reinforced the finding that these were sporadic infections. However, widespread infections detected in France [210] showed that European brown hare infected with RHDV GI.2 were common in areas that hares and rabbits live sympatrically. Detection of GI.2 related mortalities in European brown hares in Australia [211], England [212] and Scotland [213] demonstrate that the problem is recurrent. An outbreak in Germany in captive mountain hares (*Lepus timidus*) adds to the list of susceptible species and demonstrated that juvenile mountain hares could succumb to the disease [214]. Reports of substantial outbreaks in the absence of local rabbit populations in Sweden demonstrate potential hare-to-hare transmission of RHDV GI.2 [215]. Wide spread deaths in different lagomorph species have also been reported from across the USA ([216] and news article therein) including black-tailed jackrabbit (*Lepus californicus*) and desert cottontail rabbits (*Sylvilagus audubonii*) [217]. Therefore, RHDV GI.2 exhibits a broader host range than classical RHDV (GI.1) by infecting not only different rabbit species but also different hare species (*Lepus capensis mediterraneus*, *Lepus corsicanus*, *Lepus europaeus*, and *Lepus timidus*).

Different species of hare infected by RHDV GI.2, showed very similar clinical signs typical of European brown hare syndrome (EBHS): hyperaemic trachea sometimes containing uncoagulated blood, hepatitis necrosis, splenomegaly and congestion of other organs and tissues [99, 138, 206, 215].

Retrospective studies have also blurred the line defining species susceptibility to RHDV and EBHSV. Lopes and colleagues [218], identified the presence of RHDV in archival samples from Iberian hares found dead in the 1990s in Portugal with signs of an EBHS-like disease [218]. These authors demonstrated that RHDV GI.1 strains in these two cases were phylogenetically closely related to those circulating at that time and in the same areas in rabbit populations. These results would support the theory of that virus dissemination and high infection pressure in the environment could favour spillover events of infection of European brown hares with RHDV GI.2 [99, 206, 210, 211].

Analysis of RHDV GI.2 positive hares sampled in 2013 [210] have shown the existence of co-infection by EBHSV GII.1 and RHDV GI.2. This is an important issue in epidemiology and evolution, especially the potential emergence of recombinant EBHSV/RHDV GI.2 strains.

Species susceptibility to lagoviruses may be variable and this may reflect different species-specific host factors such as glycan expression for viral attachment [170]. With respect to this, several studies indicate that, as for noroviruses [219], specific binding between lagoviruses and glycans, particularly those of the HBGAs found in the upper respiratory tract and intestines of rabbits, is the first step of the viral infection [169, 220, 221]. Rabbits have different types of HBGAs and different virus strains show variable affinity to these molecules. Subtle changes in those attachment factors, e.g. through mutations, cause individual animals or even complete species can become more or less susceptible to the virus [169, 222, 223]. So, a possible explanation for overcoming species barriers could be the genetic variation of the capsid protein VP60 which alters the binding to histo-blood group antigens [223] that are considered as being important entry points for the virus [169, 221]. Other factors that could affect RHDV GI.2 infection in different species, such as concurrent subclinical infections, parasitic infestations, malnutrition or habitat detriment [138].

## 4. Viruses that infect lagomorphs Part II

A major concern regarding animal viruses are zoonotic infections. While viral zoonosis from rabbits or hares have not been commonly documented both, rabbits and hares are susceptible to infections that can infect humans including rabies, hepatitis E virus and herpes. Rabbit susceptibility to rabies was used to great benefit when Louis Pasteur endeavoured to create a rabies vaccine in the 1890s. Although susceptible to fatal rabies infection, rabbits are a spillover host. The most commonly documented source of infection in rabbits is from racoons, therefore the disease should be considered in areas where racoon rabies is endemic.

### 4.1 Hepatitis E virus

Hepatitis E virus has been detected in domestic and wild rabbits and the European brown hare in Europe and Asia, raising concerns as to whether lagomorphs maybe a reservoir for human infections [224, 225]. Although a recent study did not detect the presence of HEV in wild lagomorphs in Spain [226]. HEV belongs to the Hepeviridae virus family and has a non-segmented RNA genome comprising 3 ORFs. There are 4 genotypes, genotypes 1 and 2 cause human infections while genotpyes 3 and 4 affect wildlife species. Rabbit HEV is genotype 3. Liver or bile are the target organ for diagnosis using RT-PCR used to detect virus genome and specific ELISA to detect antibodies in sera. This pathogen should be considered when handling wild lagomorphs.

### 4.2 Herpes virus

Humans have been cited as the source of herpes virus infection in pet rabbits. With a fatal outcome the infections were the result of contact with a human with a cold sore lesion. Post-mortem analysis revealed HSV infection [227].

Naturally occurring herpes virus infections of lagomorphs have been detected in rabbit and hare species, as outlined in Chapter 1 of this book. Five putative species of Leporid herpesvirus have been described to date. Leporid herpesvirus types 1 and 3 have been isolated from *Sylvilagus floridanus* [228, 229] and Leporid herpesvirus types 2 and 4 have been isolated from *Oryctolagus cuniculus* [228–230]. Recently, during an outbreak of myxomatosis in the Iberian hare (*Lepus granatesis*) leporid herpes virus 5 was reported [73, 231]. The most studied LeHV infection is caused by LeHV-3 which causes tumor-like lesions in various organs such as liver, spleen kidney and on lymph nodes [232]. The genome sequence of LeHV-4 has been published [233], this virus caused systemic illness that began with acute ocular infections in domestic rabbits [230]. LeHV-4 has been classified as a member of the *Alphavirus* virinae subfamily, genus *Simplexvirus*. While LeHV1–3 are related to members of the *Rhadinovirus* genus in the Gammaherpes virinae they have not been approved as species (ICTV 2019 release). It has been suggested that LeHV-5 is also a gammaherpes virus [73, 231].

### 4.3 Rabbit papillomavirus

There are two species of the papillomavirus that infect lagomorphs. The first, previously termed Rabbit Papillomavirus (Shope papilloma virus or cottontail rabbit papilloma virus), predominantly infects the cottontail rabbit (*Sylvilagus floridanus*) but may also infect *Oryctolagus cuniculus*. The virus is now termed Sylvilagus floridanus papilloma virus 1 and belongs to the *Kappapapillomavirus* genus (Van Doorslaer et al.) in the family *Papillomaviridae*. The virus replicates in skin tissue causing the growth of warts on the head and neck of rabbits which can become so large that they appear to be horns and can impede animal feeding if growth occurs close to the mouth. It was the first virus to be shown to cause cancer and up to 70% of warts will lead to cancerous growths. Vaccines have been developed for use in endemic regions and the virus/rabbit infection model has served as an excellent animal model for the study of antivirals and vaccines. Thanks to such studies much is known about the molecular biology of this virus. The *Kappapapillomavirus* genus also contains a second species of virus named *Oryctolagus cuniculus* papilloma virus 1. Infection of domestic rabbits with this virus causes self-limiting oral warts or papillomas that regress without treatment. Hares have also been reported to be susceptible to papillomatosis [234].

Several viruses have been implicated as contributing factors in rabbit enteritis complex (REC; also referred to as enterocolitis or enteritis complex). REC is a complex disease of the intestine in predominantly young rabbits although the precise causes are not known it is a multifactorial disease in which bacteria, virus, parasites and environmental factors are known to be important. The presence of several RNA viruses of notable concern has recently been analysed in this regard. Rotavirus, coronavirus, astrovirus and hepatitis E virus all cause enteric disease and may potentially be of concern [235].

### 4.4 Rabbit rotavirus

Rotaviruses are an important group of segmented-genome double-stranded RNA viruses of the family *Reoviridae* that cause gasterentroitis in mammals. Rabbit or Lapine rotavirus is a group A rotavirus and have been detected in several countries. Rabbit rotavirus infection has been implicated as a factor in "multifactorial enteropathy" [236]. In Spanish rabbitries, rabbit rotaviruses have been detected and found to often be associated with other pathogens such as *Eimeria* spp., *C. spiroforme, C. perfringens, E. coli* or combinations of these agents. The authors hypothesized that damage caused by rotavirus replication in the mucosa led to a predisposition for bacterial growth and infection [237]. Predominantly found in young farmed rabbits rotaviruses have also been described in Eastern cottontail rabbits (*S. floridanus*) and hares (Snowshoe and European hares). Molecular characterisation of lapine rotavirus strains requires RT-PCR analysis with regions of the VP4, VP6 and VP7 genes being targets. Several genotypes exist due to the capacity for reassortment and genetic mutation that these segmented RNA viruses have. Detection of rabbit rotavirus in human infants have been reported [238, 239].

### 4.5 Rabbit coronavirus

Rabbit coronavirus was first described in 1961 following electron microscope detection of coronavirus particles and heart was described as the target organ [240]. Subsequently immunoelectron microscopy was used to detect Rabbit enteric coronavirus and virus was isolated [241]. Rabbit enteric CoV was detected in fecal matter during wet market surveillance in China. The characterisation of RbCoV provided the complete genome sequence (RbCoV HKU14-1 genbank accession number JN874559) has shown it to be a Betacoronavirus and reported cases detect RbCoV has also been implicated as a factor in REC. RT-PCR and RT-qPCR have been developed targeting the RdRp gene.

During the current SARS-CoV2 pandemic the potential for infection of animal hosts and the establishment of reservoirs has been of great concern. Molecular modelling studies suggest that the rabbit ACE2 molecule shares structural similarities with human ACE2 and could therefore act as a receptor for SARS-CoV2 virus entry leading to speculation that rabbit may be susceptible to infection [242]. Laboratory rabbits inoculated under experimental conditions with SARS-CoV2 were asymptomatic and low levels of infectious virus was recovered from nasal swabs up to 7 days post infection [243]. Such findings emphasize the need for strict biosafety control measures on domestic rabbit farms. At the time of writing no naturally occurring cases of SARS-CoV2 have been reported in rabbits.

### 4.6 Rabbit astrovirus

Astroviruses (family *Astroviridae*) are nonenveloped, with a single-stranded positive sense RNA genome. The discovery of astrovirus implicated in REC multifactorial disease highlighted the complexity of this disease and the need for continued surveillance [235]. Astrovirus was detected in healthy and symptomatic animals and the precise role of this virus in rabbit disease remains to be fully explored [235]. The qRT-PCR designed in this study targets the ORF1b (RNA-dependent RNA polymerase) region and provides the necessary tools for surveillance and phylogenetic analysis. Virochip coupled with metagenomic analysis allowed the identification and determination of the first full genome sequence of this agent associated with an outbreak of enterocolitis in domestic rabbits in the USA [244].

Metagenomic virome studies have shown the presence of astrovirus in rabbits from Australia and highlighted the potential for spread by insect vectors [245]. However, more analysis is required to determine the pathogenesis of this virus in rabbits.

Virome studies have also identified novel lagomorph bocaparvoviruses (genome sequence genbank accession number NC_028973) [246], picornavirus, caliciviruses amounst others [245]. The relevance of these viruses to the sanitary status of lagomorphs should be monitored. Such metagenomic studies offer novel insights into the viruses of these species and indicate the complexity of multifactorial conditions. The molecular tools that can be garnered will undoubtedly improve our understanding of the viral diseases of lagomorphs.

## 5. Concluding remarks

The detection of recent cross species transmissions of both MYXV and RHDV between lagomorph species, both sporadic and widespread and findings from the analysis of historic samples are changing our view on the species susceptible to these diseases. Rabbit and hare species are genetically and immunologically similar and, in many regions, live sympatrically, key factors when considering virus species jumps. Soon after the release of MXYV in 1950 hares were known to be susceptible to this disease, however, no large-scale infections were documented until 2018. RHDV GI.1 emerged in 1984 in the European rabbit and this was the only lagomorph species apparently affected until the emergence of RHDV GI.2 in 2010. What has driven these changes to occur is a matter for study and the effect of these species jumps on lagomorph populations has yet to be seen. Thanks to the great effort of historic surveillance studies and careful sample archiving, the molecular evolution of these viruses is being discovered.

Metagenomic studies have also identified novel lagomorph viruses. Through such studies and continued surveillance therapeutics to lesser known lagomorph viruses and a better understanding of animal health will ensue. We may now be entering a new era in the study of the viruses that infect lagomorphs which will further our understanding on the complexity of virus-host relationship.

## Acknowledgements

FP received funding from the Spanish Ministry for Science Innovation and Universities (grant number AGL2017-83395-R) and INTERCUN and the Ayuntamiento Ribera de Arriba, Asturias, Spain. ICS received a pre-doctoral grant from the Spanish Ministry for Science Innovation and Universities (reference number PRE2018-087157). KPD and JMMA acknowledge funding from the Spanish Ministry for Science Innovation and Universities (reference number PID2020-120349RB-100).

## Conflict of interest

"The authors declare no conflict of interest."

## Author details

Kevin P. Dalton[1*], Ana Podadera[1], José Manuel Martin Alonso[1], Inés Calonge Sanz[1], Ángel Luis Álvarez Rodríguez[1], Rosa Casais[2] and Francisco Parra[1]

1 Department of Biochemistry and Molecular Biology, University Institute of Biotechnology of Asturias, University of Oviedo, Oviedo, Spain

2 Centro de Biotecnología Animal, Servicio Regional de Investigación y Desarrollo Agroalimentario (SERIDA), Deva, Asturias, Spain

*Address all correspondence to: daltonkevin@uniovi.es

## References

[1] Sanarelli G. No Title. Centr Bakt Abt 1. 1898;23:865.

[2] Kerr PJ. Myxomatosis in Australia and Europe: A model for emerging infectious diseases. Antiviral Res. 2012;93(3):387-415.

[3] Kerr PJ, Best SM. Myxoma virus in rabbits. Rev Sci Tech. 1998;17(1):256-68.

[4] Fenner F, Ratcliffe FN. Myxomatosis. Cambridge: Cambridge University Press; 1965.

[5] Kerr PJ, Liu J, Cattadori I, Ghedin E, Read AF, Holmes EC. Myxoma virus and the leporipoxviruses: An evolutionary paradigm. Viruses. 2015;7(3):1020-61.

[6] Alves JM, Carneiro M, Cheng JY, de Matos AL, Rahman MM, Loog L, et al. Parallel adaptation of rabbit populations to myxoma virus. Science (80- ). 2019;363(6433):1319-26.

[7] Mead-Briggs AR, Vaughan JA. The differential transmissibility of myxoma virus strains of differing virulence grades by the rabbit flea Spilopsyllus cuniculi (Dale). J Hyg (Lond). 1975;75(2):237-47.

[8] Saint KM, French N, Kerr P. Genetic variation in Australian isolates of myxoma virus: An evolutionary and epidemiological study. Arch Virol. 2001;146(6):1105-23.

[9] Cameron C, Hota-Mitchell S, Chen L, Barrett J, Cao JX, Macaulay C, et al. The complete DNA sequence of myxoma virus. Virology. 1999;264(2):298-318.

[10] Spiesschaert B, McFadden G, Hermans K, Nauwynck H, Van De Walle GR. The current status and future directions of myxoma virus, a master in immune evasion. Vet Res. 2011;42(1):1-18.

[11] Camus-Bouclainville C, Gretillat M, Py R, Gelfi J, Guérin JL, Bertagnoli S. Genome Sequence of SG33 strain and recombination between wild-type and vaccine myxoma viruses. Emerg Infect Dis. 2011;17(4):633-8.

[12] Upton C, Slack S, Hunter AL, Ehlers A, Roper RL. Poxvirus Orthologous Clusters: toward Defining the Minimum Essential Poxvirus Genome. J Virol. 2003;77(13):7590-600.

[13] Fenner F, Woodroofe G. PROTECTION OF LABORATORY RABBITS AGAINST MYXOMATOSIS BY VACCINATION WITH FIBROMA VIRUS. Aust J Exp Biol Med Sci. 1954;32(5):653-68.

[14] Shope RE. A FILTRABLE VIRUS CAUSING A TUMOR-LIKE CONDITION IN RABBITS AND ITS RELATIONSHIP TO VIRUS MYXOMATOSUM. 1932;(1).

[15] The S, Diseases I, Dec N. Attenuation of the Myxoma Virus and Use of the Living Attenuated Virus as an Immunizing Agent for Myxomatosis Author ( s ): J. K. Saito , D. G. McKercher and G. Castrucci Published by : Oxford University Press Stable URL : https://www.jstor.org/stabl. 2020;114(5):417-28.

[16] McKercher D, Saito J. AN ATTENUATED LIVE VIRUS VACCINE FOR MYXOMATOSIS. Nature. 1964;202:933-4.

[17] Arguello Villares J. La mixomatosis hoy y su profilaxis. Cunicultura. 1984;Agosto:121-7.

[18] Jeklova E, Leva L, Matiasovic J, Kovarcik K, Kudlackova H, Nevorankova Z, et al. Characterisation of immunosuppression in rabbits afterinfection with myxoma virus. Vet Microbiol. 2008;129:117-30.

[19] Manev I, Genova K, Lavazza A, Capucci L. Humoral immune response to different routes of myxomatosis vaccine application. World Rabbit Sci. 2018;26(2):149-54.

[20] Cavadini P, Botti G, Barbieri I, Lavazza A, Capucci L. Molecular characterization of SG33 and Borghi vaccines used against myxomatosis. Vaccine. 2010;28(33):5414-20.

[21] Braun C, Thürmer A, Daniel R, Schultz A-K, Bulla I, Schirrmeier H, et al. Genetic Variability of Myxoma Virus Genomes. J Virol. 2017;91(4).

[22] Saito J, McKercher D, Castrucci G. Attenuation of the Myxoma Virus and Use of the Living Attenuated Virus as an Immunizing Agent for Myxomatosis. J Infect Dis. 1964;114(5):417-28.

[23] Opgenorth A, Graham K, Nation N, Strayer D, McFadden G. Deletion analysis of two tandemly arranged virulence genes in myxoma virus, M11L and myxoma growth factor. J Virol. 1992;66(0022-538X SB-M SB-X):4720-31.

[24] Lamb S, Rahman M, McFadden G. Recombinant myxoma virus lacking all poxvirus ankyrin-repeat proteins stimulates multiple cellular anti-viral pathways and exhibits a severe decrease in virulence. Virology. 2014;(1):134-45.

[25] Bertagnoli S, Gelfi J, Le Gall G, Boilletot E, Vautherot JF, Rasschaert D, et al. Protection against myxomatosis and rabbit viral hemorrhagic disease with recombinant myxoma viruses expressing rabbit hemorrhagic disease virus capsid protein. J Virol. 1996;70(8):5061-6.

[26] Spibey N, McCabe VJ, Greenwood NM, Jack SC, Sutton D, Van Der Waart L. Novel bivalent vectored vaccine for control of myxomatosis and rabbit haemorrhagic disease. Vet Rec. 2012;170(12):309.

[27] Torres J, Ramirez M, Morales M, Barcena J, Vazquez B, Espuña E, et al. Safety evaluation of a recombinant myxoma-RHDV virus inducing horizontal transmissible protection against myxomatosis and rabbit haemorrhagic disease. Vaccine. 2001;19:174-82.

[28] Bárcena J, Pagès-Manté A, March R, Morales M, Ramírez MA, Sánchez-Vizcaíno JM, et al. Isolation of an attenuated myxoma virus field strain that can confer protection against myxomatosis on contacts of vaccinates. Arch Virol. 2000;145(4):759-71.

[29] McCabe VJ, Tarpey I, Spibey N. Vaccination of cats with an attenuated recombinant myxoma virus expressing feline calicivirus capsid protein. Vaccine. 2002;20(19-20):2454-62.

[30] Pignolet B, Boullier S, Gelfi J, Bozzetti M, Russo P, Foulon E, et al. Safety and immunogenicity of myxoma virus as a new viral vector for small ruminants. J Gen Virol. 2008;89(6):1371-9.

[31] Top S, Foucras G, Deplanche M, Rives G, Calvalido J, Comtet L, et al. Myxomavirus as a vector for the immunisation of sheep: Protection study against challenge with bluetongue virus. Vaccine. 2012;30(9):1609-16.

[32] Rahman MM, McFadden G. Oncolytic Virotherapy with Myxoma Virus. J Clin Med. 2020;9(1):171.

[33] Dalton KP, Nicieza I, de Llano D, Gullón J, Inza M, Petralanda M, et al. Vaccine breaks: Outbreaks of myxomatosis on Spanish commercial rabbit farms. Vet Microbiol. 2015;178(3-4):208-16.

[34] Kerr PJ, Rogers MB, Fitch A, Depasse J V, Cattadori IM, Twaddle AC, et al. Genome scale evolution of myxoma virus reveals host-pathogen adaptation and rapid geographic spread. J Virol. 2013;87(23):12900-15.

[35] Kerr PJ, Cattadori IM, Rogers MB, Fitch A, Geber A, Liu J, et al. Genomic and phenotypic characterization of myxoma virus from Great Britain reveals multiple evolutionary pathways distinct from those in Australia. PLoS Pathog. 2017;13(3):1-23.

[36] Kerr PJ, Ghedin E, DePasse J V., Fitch A, Cattadori IM, Hudson PJ, et al. Evolutionary History and Attenuation of Myxoma Virus on Two Continents. PLoS Pathog. 2012;8(10).

[37] Kerr PJ, Eden JS, Di Giallonardo F, Peacock D, Liu J, Strive T, et al. Punctuated evolution of myxoma virus: Rapid and disjunct evolution of a recent viral lineage in Australia. bioRxiv. 2018;93(8):1-18.

[38] Mutze G, Bird P, Kovaliski J, Peacock D, Jennings S, Cooke B. Emerging epidemiological patterns in rabbit haemorrhagic disease, its interaction with myxomatosis, and their effects on rabbit populations in South Australia. Wildl Res. 2002;29(6):577-90.

[39] Liu J, Cattadori IM, Sim DG, Eden J-S, Holmes EC, Read AF, et al. Reverse Engineering Field Isolates of Myxoma Virus Demonstrates that Some Gene Disruptions or Losses of Function Do Not Explain Virulence Changes Observed in the Field. J Virol. 2017;91(20):1-18.

[40] Mossman K, Lee SF, Barry M, Boshkov L, McFadden G. Disruption of M-T5, a novel myxoma virus gene member of poxvirus host range superfamily, results in dramatic attenuation of myxomatosis in infected European rabbits. J Virol. 1996;70(7):4394-410.

[41] Mansouri M, Bartee E, Gouveia K, Hovey Nerenberg BT, Barrett J, Thomas L, et al. The PHD/LAP-Domain Protein M153R of Myxomavirus Is a Ubiquitin Ligase That Induces the Rapid Internalization and Lysosomal Destruction of CD4. J Virol. 2003;77(2):1427-40.

[42] Guerin J-L, Gelfi J, Boullier S, Delverdier M, Bellanger F-A, Bertagnoli S, et al. Myxoma Virus Leukemia-Associated Protein Is Responsible for Major Histocompatibility Complex Class I and Fas-CD95 Down-Regulation and Defines Scrapins, a New Group of Surface Cellular Receptor Abductor Proteins. J Virol. 2002;76(6):2912-23.

[43] Camus-Bouclainville C, Fiette L, Bouchiha S, Pignolet B, Counor D, Filipe C, et al. A virulence factor of myxoma virus colocalizes with NF-kappaB in the nucleus and interferes with inflammation. J Virol. 2004;78(5):2510-6.

[44] Blanié S, Gelfi J, Bertagnoli S, Camus-Bouclainville C. MNF, an ankyrin repeat protein of myxoma virus, is part of a native cellular SCF complex during viral infection. Virol J. 2010;7:1-5.

[45] Bárcena J, Morales M, Vázquez B, Boga JA, Parra F, Lucientes J, et al. Horizontal transmissible protection against myxomatosis and rabbit hemorrhagic disease by using a recombinant myxoma virus. J Virol. 2000;74(3):1114-23.

[46] Morales M, Ramírez M a, Cano MJ, Párraga M, Castilla J, Pérez-Ordoyo LI, et al. Genome comparison of a nonpathogenic myxoma virus field strain with its ancestor, the virulent Lausanne strain. J Virol. 2009;83(5):2397-403.

[47] Dalton KP, Nicieza I, Baragaño A, Martín-Alonso J, Parra F. Molecular characterisation of virulence graded field isolates of myxoma virus. Virol J. 2010;7:49.

[48] Flowerdew JR, Trout RC, Ross J. Myxomatosis: population dynamics of

rabbits (*Oryctolagus cuniculus* Linnaeus, 1758) and ecological effects in the United Kingdom. Rev Sci Tech. 1992;11(4):1109-13.

[49] Moore N. RABBITS, BUZZARDS AND HARES. TWO STUDIES ON THE INDIRECT EFFECTS OF MYXOMA TOSIS. Rev D'Écologie La Terre la vie. 1956;3-4(1):1-8.

[50] Sumption K, Flowerdew J. The ecological effects of the decline in Rabbits (*Oryctolagus cuniculus* L.) due to myxomatosis. Mamm Rev. 1985;15(4):151-86.

[51] Valverde JA. Estructura de una comunidad mediterranea de vertebrados terrestres. Monogr Cienc Mod. 1967;76:1-236.

[52] Delibes-Mateos M, Redpath SM, Angulo E, Ferreras P, Villafuerte R. Rabbits as a keystone species in southern Europe. Biol Conserv. 2007;137(1):149-56.

[53] Kerr PJ. An ELISA for epidemiological studies of Myxomatosis: Persistence of antibodies to myxoma virus in European rabbits (*Oryctolagus cuniculus*). Wildl Res. 1997;24:53-65.

[54] Louzis C, Arthur CP, Louzis C. A review of myxomatosis among rabbits in France. Rev sci tech Off int Epiz. 1988;7(4):959-76.

[55] Rosell JM, de la Fuente LF, Parra F, Dalton KP, Sáiz JIB, de Rozas AP, et al. Myxomatosis and rabbit haemorrhagic disease: A 30-year study of the occurrence on commercial farms in Spain. Animals. 2019;9(10).

[56] Kwit E, Rzeżutka A. Molecular methods in detection and epidemiologic studies of rabbit and hare viruses: a review. J Vet Diagnostic Investig. 2019;31(4):497-508.

[57] OIE. Myxomatosis. In: OIE Terrestrial Manual. 2018.

[58] Albini S, Sigrist B, Güttinger R, Schelling C, Hoop RK, Vögtlin A. Development and validation of a Myxoma virus real-time polymerase chain reaction assay. J Vet Diagnostic Investig. 2012;24(1):135-7.

[59] Belsham GJ, Polacek C, Breum SØ, Larsen LE, Bøtner A. Detection of myxoma viruses encoding a defective M135R gene from clinical cases of myxomatosis; possible implications for the role of the M135R protein as a virulence factor. Virol J. 2010;7(7).

[60] Duarte MD, Barros SC, Henriques AM, Fagulha MT, Ramos F, Luís T, et al. Development and validation of a real time PCR for the detection of myxoma virus based on the diploid gene M000.5L/R. J Virol Methods. 2013 Dec;196:219-24.

[61] Farsang a, Makranszki L, Dobos-Kovács M, Virág G, Fábián K, Barna T, et al. Occurrence of atypical myxomatosis in Central Europe: clinical and virological examinations. Acta Vet Hung. 2003;51(4):493-501.

[62] Alda F, Gaitero T, Suárez M, Doadrio I, Suárez M, Doadrio I. Molecular characterisation and recent evolution of myxoma virus in Spain. Arch Virol. 2009;154(10):1659-70.

[63] Muller A., Silva E, Abrantes J, Esteves PJ, Ferreira PG, Carvalheira JC, et al. Partial sequencing of recent Portuguese myxoma virus field isolates exhibits a high degree of genetic stability. Vet Microbiol. 2010;140(1-2):161-6.

[64] Esteban DJ, Hutchinson AP. Genes in the terminal regions of orthopoxvirus genomes experience adaptive molecular evolution. BMC Genomics. 2011;12.

[65] Dalton KP, Ringleb F, Martín Alonso JM, Parra F. Rapid identification of myxoma virus variants by long-range PCR and restriction fragment length polymorphism analysis. J Virol Methods. 2009;161(2):284-8.

[66] Gelfi J, Chantal J, Phong TT, Py R, Boucraut-Baralon C. Development of an ELISA for detection of myxoma virus-specific rabbit antibodies: test evaluation for diagnostic applications on vaccinated and wild rabbit sera. J Vet Diagn Invest. 1999;11(3):240-5.

[67] Lavazza A, Graziani M, Tranquillo VM, Botti G, Palotta C, Cerioli M, et al. Serological evaluation of the immunity induced in commercial rabbits by vaccination for myxomatosis and RHD. Proc 8th World Rabbit Congr Puebla. 2004;569-75.

[68] Barlow A, Lawrence K, Everest D, Dastjerdi A, Finnegan C, Steinbach F. Confirmation of myxomatosis in a European brown hare in Great Britain. Vet Rec. 2014;175(3):75-6.

[69] García-Bocanegra I, Camacho-Sillero L, Risalde M, Dalton KP, Caballero-Gómez J, Agüero M, et al. First outbreak of myxomatosis in Iberian hares (*Lepus granatensis*). Transbound Emerg Dis. 2019;66(6):2204-8.

[70] Dalton KP, Martín JM, Nicieza I, Podadera A, de Llano D, Casais R, et al. Myxoma virus jumps species to the Iberian hare. Transbound Emerg Dis. 2019;66(6):2218-26.

[71] Águeda-Pinto A, de Matos AL, Abrantes M, Kraberger S, Risalde MA, Gortázar C, et al. Genetic characterization of a recombinant myxoma virus leap into the Iberian hare (*Lepus granatensis*). bioRxiv. 2019;1-16.

[72] García-Bocanegra I, Camacho-Sillero L, Caballero-Gómez J, Agüero M, Gómez-Guillamón F, Manuel Ruiz-Casas J, et al. Monitoring of emerging myxoma virus epidemics in Iberian hares (*Lepus granatensis*) in Spain, 2018-2020. Transbound Emerg Dis. 2020;(May):1-8.

[73] Abade dos Santos FA, Carvalho CL, Pinto A, Rai R, Monteiro M, Carvalho P, et al. Detection of recombinant hare myxoma virus in wild rabbits (*Oryctolagus cuniculus algirus*). Viruses. 2020;12(10):1-12.

[74] Qin L, Evans DH. Genome Scale Patterns of Recombination between Coinfecting Vaccinia Viruses. J Virol. 2014;88(10):5277-86.

[75] Gershon PD, Kitching RP, Hammond JM, Black DN. Poxvirus genetic recombination during natural virus transmission. J Gen Virol. 1989;70(2):485-9.

[76] Yao X-D, Evans DH. High-Frequency Genetic Recombination and Reactivation of Orthopoxviruses from DNA Fragments Transfected into Leporipoxvirus-Infected Cells. J Virol. 2003;77(13):7281-90.

[77] Cancellotti FM, Renzi M. Epidemiology and current situation of viral haemorrhagic disease of rabbits and the European brown hare syndrome in Italy. Rev Sci Tech. 1991;10(2):409-22.

[78] Abrantes J, Loo W Van Der, Pendu J Le, Esteves PJ, van der Loo W, Le Pendu J, et al. Rabbit haemorrhagic disease (RHD) and rabbit haemorrhagic disease virus (RHDV): A review. Vet Res. 2012;43(1):12.

[79] Chasey D. Rabbit haemorrhagic disease: the new scourge of *Oryctolagus cuniculus*. Lab Anim. 1997;31:33-44.

[80] Argüello Villares J, Llanos Pellitero A, Pérez-Ordoyo Garcia L. Enfermedad vírica hemorragica canejo en España. Rev Med Vet. 1988;5(12):645-50.

[81] Villafuerte R, Calvete C, Blanco J, Lucientes J. Incidence of viral hemorrhagic disease in wild rabbit populations in Spain. Mammalia. 1995;59(4):651-9.

[82] Calvete C, Estrada R, Villafuerte R, Osácar JJ, Lucientes J. Epidemiology of viral haemorrhagic disease and myxomatosis in a free-living population of wild rabbits. Vet Rec. 2002;150(25):776-82.

[83] Delibes-Mateos M, Ramírez E, Ferreras P, Villafuerte R. Translocations as a risk for the conservation of European wild rabbit *Oryctolagus cuniculus* lineages. Oryx. 2008;42(2):259-64.

[84] Cooke B, Fenner F. Rabbit haemorrhagic disease and the biological control of wild rabbits, *Oryctolagus cuniculus*, in Australia and New Zealand. Wildl Res. 2003;29:689-706.

[85] Arguello Villares J. Viral haemorrhagic disease of rabbits: vaccination and immune response. Rev sei tech Off int Epiz,. 1991;10(2):471-80.

[86] Smid B, Valíček L, Rodak L, Stepanek J, Jurak E. Rabbit haemorrhagic disease: an investigation of some properties of the virus and evaluation of an inactivated vaccine. Vet Microbiol. 1991;26(1-2):77-85.

[87] Ohlinger V, Thiel H. Identification of the viral haemorrhagic disease virus of rabbits as a calicivirus. Rev sci tech Off int Epiz,. 1991;10(2):311-23.

[88] Morisse J, Le Gall G, Boilletot E. Hepatitis of viral origin in Leporidae: introduction and aetiological hypotheses. Rev sci tech Off int Epiz,. 1991;10(2):283-95.

[89] Asgari S, Hardy JR., Sinclair RG, Cooke BD. Field evidence for mechanical transmission of rabbit haemorrhagic disease virus (RHDV) by flies (Diptera: Calliphoridae) among wild rabbits in Australia. Virus Res. 1998 Apr;54(2):123-32.

[90] McColl KA, Merchant JC, Hardy J, Cooke BD, Robinson A, Westbury HA. Evidence for insect transmission of rabbit haemorrhagic disease virus. Epidemiol Infect. 2002;129(3):655-63.

[91] Le Gall-Reculé G, Zwingelstein F, Boucher S, Le Normand B, Plassiart G, Portejoie Y, et al. Detection of a new variant of rabbit haemorrhagic disease virus in France. Vet Rec. 2011;168(5):137-8.

[92] Dalton KP, Nicieza I, Balseiro A, Muguerza MA, Rosell JMJ, Casais R, et al. Variant rabbit hemorrhagic disease virus in young rabbits, Spain. Emerg Infect Dis. 2012;18(12):2009-12.

[93] OIE. Rabbit Haemorrhagic Disease: Aetiology Epidemiology Diagnosis Prevention and Control References. 2019;1-7. Available from: https://www.oie.int/fileadmin/Home/eng/Animal_Health_in_the_World/docs/pdf/Disease_cards/RHD.pdf

[94] Le Pendu J, Abrantes J, Bertagnoli S, Guitton J-SS, Le Gall-Reculé G, Lopes AM, et al. Proposal for a unified classification system and nomenclature of lagoviruses. J Gen Virol. 2017 Jul 1;98(7):1658-66.

[95] Le, Gall-Reculé G, Lavazza A, Marchandeau S, Bertagnoli S, Zwingelstein F, Cavadini P, et al. Emergence of a new lagovirus related to Rabbit Haemorrhagic Disease Virus. Vet Res. 2013;44:81.

[96] Abrantes J, Lopes AM, Dalton KP, Melo P, Correia J, Ramada M, et al. New variant of rabbit hemorrhagic disease virus, Portugal, 2012-2013. Emerg Infect Dis. 2013;19(11):1900-2.

[97] Dalton KP, Nicieza I, Abrantes J, Esteves P, Parra F. Spread of new variant

RHDV in domestic rabbits on the Iberian Peninsula. Vet Microbiol. 2014;169(1-2):67-73.

[98] Calvete C, Mendoza M, Alcaraz A, Sarto MP, Jiménez-de-Bagüéss MP, Calvo AJ, et al. Rabbit haemorrhagic disease: Cross-protection and comparative pathogenicity of GI.2/RHDV2/b and GI.1b/RHDV lagoviruses in a challenge trial. Vet Microbiol. 2018;219(April):87-95.

[99] Puggioni G, Cavadini P, Maestrale C, Scivoli R, Botti G, Ligios C, et al. The new French 2010 Rabbit Hemorrhagic Disease Virus causes an RHD-like disease in the Sardinian Cape hare (*Lepus capensis mediterraneus*). Vet Res. 2013;44(1):1-7.

[100] Mahar JE, Hall RN, Peacock D, Kovaliski J, Piper M, Mourant R, et al. Rabbit Hemorrhagic Disease Virus 2 (RHDV2; GI.2) Is Replacing Endemic Strains of RHDV in the Australian Landscape within 18 Months of Its Arrival. J Virol. 2017;

[101] Peacock D, Kovaliski J, Sinclair R, Mutze G, Lannella A, Capucci L. RHDV2 overcoming RHDV immunity in wild rabbits ( *Oryctolagus cuniculus*) in Australia. Vet Rec. 2017;180(11):280.

[102] Calvete C, Sarto P, Calvo AJ, Monroy F, Calvo JH. Could the new rabbit haemorrhagic disease virus variant (RHDVB) be fully replacing classical RHD strains in the Iberian Peninsula? World Rabbit Sci. 2014;22(1):91.

[103] Almeida T, Lopes AM, Magalhaes M, Neves F, Pinheiro A, Gonçalves D, et al. Tracking the evolution of the G1/RHDVb recombinant strains introduced from the Iberian Peninsula to the Azores islands, Portugal. Infect Genet Evol. 2015;34:307-13.

[104] Mutze G, De Preu N, Mooney T, Koerner D, McKenzie D, Sinclair R, et al. Substantial numerical decline in South Australian rabbit populations following the detection of rabbit haemorrhagic disease virus 2. Vet Rec. 2018;182(20):574.

[105] Delibes-Mateos M, Ferreira C, Carro F, Escudero MA, Gortázar C. Ecosystem effects of variant rabbit hemorrhagic disease virus, Iberian Peninsula. Emerg Infect Dis. 2014;20(12):2166-8.

[106] Mu SP, Julio IZ. Diseño de un virus recombinante de Mixoma Virus ( Lausanne ) con expresión de GFP. 2017;

[107] Monterroso P, Garrote G, Serronha A, Santos E, Delibes-Mateos M, Abrantes J, et al. Disease-mediated bottom-up regulation: An emergent virus affects a keystone prey, and alters the dynamics of trophic webs. Sci Rep. 2016;6(October):1-9.

[108] Rodak L, Smid B, Valicek L, Vesely T, Stepanek J, Hampl J, et al. Enzyme-linked immunosorbent assay of antibodies to rabbit haemorrhagic disease virus and determination of its major structural proteins. J Gen Virol. 1990;71(5):1075-80.

[109] Capucci L, Fusi P, Lavazza a, Pacciarini ML, Rossi C. Detection and preliminary characterization of a new rabbit calicivirus related to rabbit hemorrhagic disease virus but nonpathogenic. J Virol. 1996;70(12):8614-23.

[110] Strive T, Wright JD, Robinson a. J. Identification and partial characterisation of a new lagovirus in Australian wild rabbits. Virology. 2009;384(1):97-105.

[111] Le Gall-Reculé G, Zwingelstein F, Fages MP, Bertagnoli S, Gelfi J, Aubineau J, et al. Characterisation of a non-pathogenic and non-protective

infectious rabbit lagovirus related to RHDV. Virology. 2011;410(2):395-402.

[112] Cooke BD. Rabbit haemorrhagic disease: Field epidemiology and the management of wild rabbit populations. OIE Rev Sci Tech. 2002;21(2):347-58.

[113] Strive T, Elsworth P, Liu J, Wright JD, Kovaliski J, Capucci L. The non-pathogenic Australian rabbit calicivirus RCV-A1 provides temporal and partial cross protection to lethal Rabbit Haemorrhagic Disease Virus infection which is not dependent on antibody titres. Vet Res. 2013;44(1):1.

[114] Gavier-Widén D, Mörner T. Epidemiology and diagnosis of the European brown hare syndrome in Scandinavian countries: a review. Rev Sci Tech. 1991;10(2):453-8.

[115] Billinis C, Psychas V, Tontis DK, Spyrou V, Birtsas PK, Sofia M, et al. European brown hare syndrome in wild European brown hares from Greece. J Wildl Dis. 2005;41(4):783-6.

[116] Chasey D, Duff P. European brown hare syndrome and associated virus particles in the UK. Vet Rec. 1990;126(25):623-4.

[117] Frölich K, Fickel J, Ludwig A, Lieckfeldt D, Streich WJ, Jurčík R, et al. New variants of European brown hare syndrome virus strains in free-ranging European brown hares (*Lepus europaeus*) from Slovakia. J Wildl Dis. 2007;43(1):89-96.

[118] Frölich K, Graf Kujawski OEJ, Rudolph M, Ronsholt L, Speck S. European brown hare syndrome virus in free-ranging European brown hares from Argentina. J Wildl Dis. 2003;39(1):121-4.

[119] Frölich K, Meyer HD, Pielowski Z, Ronsholt L, von Seck-Lanzendorf S, Stolte M. European in Poland Syndrome in Free-Ranging Hares in Poland. J Wildl Dis. 1996;32(2).

[120] Frölich K, Wisser J, Schmüser H, Fehlberg U, Neubauer H, Grunow R, et al. Epizootiologic and ecologic investigations of European brown hares (*Lepus europaeus*) in selected populations from Schleswig-Holstein, Germany. J Wildl Dis. 2003;39(4):751-61.

[121] Zanni ML, Benassi MC, Scicluna MT, Lavazza A, Capucci L. Clinical evolution and diagnosis of an outbreak of European brown hare syndrome in hares reared in captivity. Rev Sci Tech. 1993;12(3):931-40.

[122] Lopes AM, Capucci L, Gavier-Widén D, Le Gall-Reculé G, Brocchi E, Barbieri I, et al. Molecular evolution and antigenic variation of European brown hare syndrome virus (EBHSV). Virology. 2014;468-470:104-12.

[123] Bascuñana CR, Nowotny N, Belák S. Detection and differentiation of rabbit hemorrhagic disease and European brown hare syndrome viruses by amplification of VP60 genomic sequences from fresh and fixed tissue specimens. J Clin Microbiol. 1997;35(10):2492-5.

[124] Gavier-Widén D, Mörner T. Descriptive epizootiological study of European brown hare syndrome in Sweden. J Wildl Dis. 1993;29(1):15-20.

[125] Lavazza A, Cavadini P, Barbieri I, Tizzani P, Pinheiro A, Abrantes J, et al. Field and experimental data indicate that the eastern cottontail (Sylvilagus floridanus) is susceptible to infection with European brown hare syndrome (EBHS) virus and not with rabbit haemorrhagic disease (RHD) virus. Vet Res. 2015;46(1):1-10.

[126] Lavazza A, Scicluna MT, Capucci L. Susceptibility of Hares and

Rabbits to the European Brown Hare Syndrome Virus (EBHSV) and Rabbit Haemorrhagic Disease Virus (RHDV) under Experimental Conditions*. J Vet Med Ser B. 1996 Jan 12; 43(1-10):401-10.

[127] Marcato PS, Benazzi C, Vecchi G, Galeotti M, Salda L Della, Sarli G, et al. Clinical and pathological features of viral haemorrhagic disease of rabbits and the European brown hare syndrome. Vol. 10, Rev. Sci. tech. Off. int. Epiz. 1991.

[128] Fuchs A, Weissenbock H. Comparative histopathological study of rabbit haemorrhagic disease (RHD) and European brown hare syndrome (EBHS). J Comp Pathol. 1992;107(1):103-13.

[129] Cavadini P, Molinari S, Merzoni F, Vismarra A, Posautz A, Alzaga Gil V, et al. Widespread occurrence of the non-pathogenic hare calicivirus (HaCV Lagovirus GII.2) in captive-reared and free-living wild hares in Europe. Transbound Emerg Dis. 2020;(June):1-10.

[130] Droillard C, Lemaitre E, Chatel M, Guitton J-S, Marchandeau S, Eterradossi N, et al. First Complete Genome Sequence of a Hare Calicivirus Strain Isolated from *Lepus europaeus*. Microbiol Resour Announc. 2018;7(22):4-5.

[131] Droillard C, Lemaitre E, Chatel M, Quéméner A, Briand FX, Guitton JS, et al. Genetic diversity and evolution of Hare Calicivirus (HaCV), a recently identified lagovirus from *Lepus europaeus*. Infect Genet Evol. 2020;82(March):104310.

[132] Mahar JE, Hall RN, Shi M, Mourant R, Huang N, Strive T, et al. The discovery of three new hare lagoviruses reveals unexplored viral diversity in this genus. Virus Evol. 2019;5(1):1-11.

[133] Cavadini P, Molinari S, Pezzoni G, Chiari M, Brocchi E, Lavazza A, et al. Identification of a new non-pathogenic lagovirus in Lepus europeaus. In: 10th International Congress for Veterinary Virology, 9th Annual Epizone Meeting: "Changing Viruses in a Changing World." 2015. p. 76-7.

[134] Capucci L, Scicluna MT, Lavazza A. Diagnosis of viral haemorrhagic disease of rabbits and the European brown hare syndrome. Rev Sci Tech. 1991;10(2):347-70.

[135] Wirblich C, Meyers G, Ohlinger VF, Capucci L, Eskens U, Haas B, et al. European brown hare syndrome virus: relationship to rabbit hemorrhagic disease virus and other caliciviruses. J Virol. 1994 Aug;68(8):5164-73.

[136] Le Gall G, Huguet S, Vende P, Vautherot JF, Rasschaert D. European brown hare syndrome virus: Molecular cloning and sequencing of the genome. J Gen Virol. 1996;77(8):1693-7.

[137] Laurent S, Vautherot JF, Le Gall G, Madelaine MF, Rasschaert D. Structural, antigenic and immunogenic relationships between European brown hare syndrome virus and rabbit haemorrhagic disease virus. J Gen Virol. 1997;78(11):2803-11.

[138] Velarde R, Cavadini P, Neimanis A, Cabezón O, Chiari M, Gaffuri A, et al. Spillover Events of Infection of Brown Hares (*Lepus europaeus*) with Rabbit Haemorrhagic Disease Type 2 Virus (RHDV2) Caused Sporadic Cases of an European Brown Hare Syndrome-Like Disease in Italy and Spain. Transbound Emerg Dis. 2017;64(6):1750-61.

[139] Szillat KP, Höper D, Beer M, König P. Full-genome sequencing of German rabbit haemorrhagic disease virus uncovers recombination between RHDV (GI.2) and EBHSV (GII.1). Virus Evol. 2020;6(2):1-11.

[140] Meyers G, Wirblich C, Thiel H. Rabbit hemorrhagic disease virus--molecular cloning and nucleotide sequencing of a calicivirus genome. Virology. 1991;184(2):664-76.

[141] Parra F, Prieto M. Purification and characterization of a calicivirus as the causative agent of a lethal hemorrhagic disease in rabbits. J Virol. 1990 Aug;64(8):4013-5.

[142] Meyers G. Translation of the Minor Capsid Protein of a Calicivirus Is Initiated by a Novel Termination-dependent Reinitiation Mechanism. J Biol Chem. 2003;278(36):34051-60.

[143] Meyers G, Wirblich C, Thiel HJ. Rabbit hemorrhagic disease virus--molecular cloning and nucleotide sequencing of a calicivirus genome. Virology. 1991 Oct;184(2):664-76.

[144] Meyers G, Wirblich C, Thiel HJ, Thumfart JO. Rabbit hemorrhagic disease virus: genome organization and polyprotein processing of a calicivirus studied after transient expression of cDNA constructs. Virology. 2000;276(2):349-63.

[145] Wirblich C, Thiel HJ, Meyers G. Genetic map of the calicivirus rabbit hemorrhagic disease virus as deduced from in vitro translation studies. J Virol. 1996;70(11):7974-83.

[146] Machín A, Martín Alonso J, Parra F. Identification of the Amino Acid Residue Involved in Rabbit Hemorrhagic Disease Virus VPg Uridylylation. J Biol Chem. 2001;276(30):27787-92.

[147] Zhu J, Wang B, Miao Q, Tan Y, Li C, Chen Z, et al. Viral Genome-Linked Protein (VPg) Is Essential for Translation Initiation of Rabbit Hemorrhagic Disease Virus (RHDV). PLoS One. 2015;10(11):e0143467.

[148] Martín Alonso JM, Casais R, Boga JA, Parra F. Processing of rabbit hemorrhagic disease virus polyprotein. J Virol. 1996;70(2):1261-5.

[149] Wennesz R, Luttermann C, Kreher F, Meyers G. Structure–function relationship in the 'termination upstream ribosomal binding site' of the calicivirus rabbit hemorrhagic disease virus. Nucleic Acids Res. 2019;47(4):1920-34.

[150] Subba-Reddy C V, Goodfellow I, Kao CC. VPg-primed RNA synthesis of norovirus RNA-dependent RNA polymerases by using a novel cell-based assay. J Virol. 2011;85(24):13027-37.

[151] König M, Thiel H-J, Meyers G. Detection of Viral Proteins after Infection of Cultured Hepatocytes with Rabbit Hemorrhagic Disease Virus. J Virol. 1998;72(5):4492-7.

[152] Vázquez AL, Martín Alonso JM, Casais R, Boga JA, Parra F. Expression of Enzymatically Active Rabbit Hemorrhagic Disease Virus RNA-Dependent RNA Polymerase in *Escherichia coli*. J Virol. 1998;72(4):2999-3004.

[153] López Vázquez A, Martín Alonso JM, Parra F. Characterisation of the RNA-dependent RNA polymerase from Rabbit hemorrhagic disease virus produced in *Escherichia coli*. Arch Virol. 2001;146(1):59-69.

[154] Machín A, Martín Alonso J, Dalton KP, Parra F. Functional differences between precursor and mature forms of the RNA-dependent RNA polymerase from rabbit hemorrhagic disease virus. J Gen Virol. 2009;90(9):2114-8.

[155] Boniotti B, Wirblich C, Sibilia M, Meyers G, Thiel HJ, Rossi C. Identification and characterization of a 3C-like protease from rabbit

hemorrhagic disease virus, a calicivirus. J Virol. 1994;68(10):6487-95.

[156] Joubert P, Pautigny C, Madelaine MF, Rasschaert D. Identification of a new cleavage site of the 3C-like protease of rabbit haemorrhagic disease virus. J Gen Virol. 2000;81(2):481-8.

[157] Urakova N, Frese M, Hall RN, Liu J, Matthaei M, Strive T. Expression and partial characterisation of rabbit haemorrhagic disease virus non-structural proteins. Virology. 2015;484:69-79.

[158] Ng KKS, Cherney MM, Vázquez AL, Machín Á, Martín Alonso JM, Parra F, et al. Crystal structures of active and inactive conformations of a caliciviral RNA-dependent RNA polymerase. J Biol Chem. 2002;277(2):1381-7.

[159] Nagesha HS, Wang LF, Hyatt a. D. Virus-like particles of calicivirus as epitope carriers. Arch Virol. 1999;144(12):2429-39.

[160] Pérez-Filgueira DM, Resino-Talaván P, Cubillos C, Angulo I, Barderas MG, Barcena J, et al. Development of a low-cost, insect larvae-derived recombinant subunit vaccine against RHDV. Virology. 2007;364(2):422-30.

[161] Gromadzka B, Szewczyk B, Konopa G, Fitzner A, Kęsy A. Recombinant VP60 in the form of virion-like particles as a potential vaccine against rabbit hemorrhagic disease virus. Acta Biochim Pol. 2006;53(2):371-6.

[162] Farnós O, Fernandez E, Chiong M, Parra F, Joglar M, Méndez L, et al. Biochemical and structural characterization of RHDV capsid protein variants produced in Pichia pastoris: Advantages for immunization strategies and vaccine implementation. Antiviral Res. 2009;81(1):25-36.

[163] Bárcena J, Verdaguer N, Roca R, Morales M, Angulo I, Risco C, et al. The coat protein of Rabbit hemorrhagic disease virus contains a molecular switch at the N-terminal region facing the inner surface of the capsid. Virology. 2004;322(1):118-34.

[164] Rademacher C, Krishna NR, Palcic M, Parra F, Peters T. NMR experiments reveal the molecular basis of receptor recognition by a calicivirus. J Am Chem Soc. 2008;130(11):3669-75.

[165] Wang X, Xu F, Liu J, Gao B, Liu Y, Zhai Y, et al. Atomic Model of Rabbit Hemorrhagic Disease Virus by Cryo-Electron Microscopy and Crystallography. PLoS Pathog. 2013;9(1).

[166] Capucci L, Frigoli G, Ronshold L, Lavazza A, Brocchi E, Rossi C. Antigenicity of the rabbit hemorrhagic disease virus studied by its reactivity with monoclonal antibodies. Virus Res. 1995;37:221-38.

[167] Le Gall-Reculé G, Lavazza A, Marchandeau S, Bertagnoli S, Zwingelstein F, Cavadini P, et al. Emergence of a new lagovirus related to rabbit haemorrhagic disease virus. Vet Res. 2013;44(1):1-13.

[168] Bárcena J, Guerra B, Angulo I, Gonzalez J, Valcárcel F, Mata CP, et al. Comparative analysis of rabbit hemorrhagic disease virus (RHDV) and new RHDV2 virus antigenicity, using specific virus-like particles. Vet Res. 2015 Dec 24

[169] Nyström K, Le Gall-Reculé G, Grassi P, Abrantes J, Ruvoën-Clouet N, Le Moullac-Vaidye B, et al. Histo-blood group antigens act as attachment factors of rabbit hemorrhagic disease virus infection in a virus strain-dependent manner. PLoS Pathog. 2011;7(8).

[170] Lopes AM, Breiman A, Lora M, Moullac-vaidye B Le, Galanina O,

Nyström K, et al. Host-Specific Glycans Are Correlated with Susceptibility to Infection by Lagoviruses , but Not with Their Virulence. J Virol. 2018;92(4):1-20.

[171] Valicek L, Smid B, Rodak L, Kudrna J. Electron and immunoelectron microscopy of rabbit haemorrhagic disease virus (RHDV). Arch Virol. 1990;112(3-4):271-5.

[172] Barcena J, Verdaguer N, Roca R, Morales M, Angulo I, Risco C, et al. The coat protein of Rabbit hemorrhagic disease virus contains a molecular switch at the N-terminal region facing the inner surface of the capsid. Virology. 2004;322(1):118-34.

[173] Hu Z, Tian X, Zhai Y, Xu W, Zheng D, Sun F. Cryo-electron microscopy reconstructions of two types of wild rabbit hemorrhagic disease viruses characterized the structural features of Lagovirus. Protein Cell. 2010;1(1):48-58.

[174] Thouvenin E, Laurent S, Madelaine MF, Rasschaert D, Vautherot JF, Hewat EA. Bivalent binding of a neutralising antibody to a calicivirus involves the torsional flexibility of the antibody hinge. J Mol Biol. 1997;270(2):238-46.

[175] Dalton KP, Podadera A, Granda V, Nicieza I, del Llano D, González R, et al. ELISA for detection of variant rabbit haemorrhagic disease virus RHDV2 antigen in liver extracts. J Virol Methods. 2018;251.

[176] Dalton KP, Nicieza I, Podadera A, de Llano D, Martin Alonso J, de los Toyos J, et al. Fast specific field detection of RHDVb. Transbound Emerg Dis. 2018 Feb 28;65(1):232-4.

[177] Dalton KP, Arnal JL, Benito AA, Chacón G, Martín Alonso JM, Parra F. Conventional and real time RT-PCR assays for the detection and differentiation of variant rabbit hemorrhagic disease virus (RHDVb) and its recombinants. J Virol Methods. 2018;251(October 2017):118-22.

[178] Forrester NL, Boag B, Buckley a., Moureau G, Gould E a. Co-circulation of widely disparate strains of Rabbit haemorrhagic disease virus could explain localised epidemicity in the United Kingdom. Virology. 2009;393(1):42-8.

[179] Kerr PJ, Kitchen A, Holmes EC. Origin and Phylodynamics of Rabbit Hemorrhagic Disease Virus. J Virol. 2009;83(23):12129-38.

[180] Forrester NL, Abubakr MI, Abu Elzein EME, Al-Afaleq AI, Housawi FMT, Moss SR, et al. Phylogenetic analysis of Rabbit haemorrhagic disease virus strains from the Arabian Peninsula: Did RHDV emerge simultaneously in Europe and Asia? Virology. 2006;344(2):277-82.

[181] Abu Elzein E, Al-Afaleq A. Rabbit haemorrhagic disease in Saudi Arabia. Vet Rec. 1999;144:480-1.

[182] Alda F, Gaitero T, Suárez M, Merchán T, Rocha G, Doadrio I. Evolutionary history and molecular epidemiology of rabbit haemorrhagic disease virus in the Iberian Peninsula and Western Europe. BMC Evol Biol. 2010;10(1):347.

[183] Nowotny N, Ros Bascuñana C, Ballagi-Pordány A, Gavier-Widén D, Uhlén M, Belák S. Phylogenetic analysis of rabbit haemorrhagic disease and European brown hare syndrome viruses by comparison of sequences from the capsid protein gene. Arch Virol. 1997;142(4):657-73.

[184] Esteves PJ, Abrantes J, Bertagnoli S, Cavadini P, Gavier-Widén D, Guitton JS, et al. Emergence of Pathogenicity in Lagoviruses: Evolution from Pre-existing Nonpathogenic Strains or

through a Species Jump? PLoS Pathog. 2015;11(11):56-61.

[185] Abrantes J, Droillard C, Lopes AM, Lemaitre E, Lucas P, Blanchard Y, et al. Recombination at the emergence of the pathogenic rabbit haemorrhagic disease virus Lagovirus europaeus/GI.2. Sci Rep. 2020;10(1):1-11.

[186] Merchán T, Rocha G, Alda F, Silva E, Thompson G, de Trucios SH, et al. Detection of rabbit haemorrhagic disease virus (RHDV) in nonspecific vertebrate hosts sympatric to the European wild rabbit (*Oryctolagus cuniculus*). Infect Genet Evol. 2011;11(6):1469-74.

[187] Lopes AM, Silvério D, Magalhães MJ, Areal H, Alves PC, Esteves PJ, et al. Characterization of old RHDV strains by complete genome sequencing identifies a novel genetic group. Sci Rep. 2017;7(1):1-7.

[188] Abrantes J, Lopes AM, Lemaitre E, Ahola H, Banihashem F, Droillard C, et al. Retrospective analysis shows that most rhdv gi.1 strains circulating since the late 1990s in france and sweden were recombinant gi.3p–gi.1d strains. Genes (Basel). 2020;11(8):1-13.

[189] Milton ID, Vlasak R, Nowotny N, Rodak L, Carter MJ. Genomic 3′ terminal sequence comparison of three isolates of rabbit haemorrhagic disease virus. FEMS Microbiol Lett. 1992;93(1):37-42.

[190] Le Gall G, Arnauld C, Boilletot E, Morisse JP, Rasschaert D. Molecular epidemiology of rabbit haemorrhagic disease virus outbreaks in France during 1988 to 1995. J Gen Virol. 1998;79(1):11-6.

[191] Le Gall-Reculé G, Zwingelstein F, Laurent S, De Boisséson C, Portejoie Y, Rasschaert D. Phylogenetic analysis of rabbit haemorrhagic disease virus in France between 1993 and 2000, and the characterisation of RHDV antigenic variants. Arch Virol. 2003;148(1):65-81.

[192] Capucci L, Fallacara F, Grazioli S, Lavazza A, Pacciarini ML, Brocchi E. A further step in the evolution of rabbit hemorrhagic disease virus: The appearance of the first consistent antigenic variant. Virus Res. 1998;58(1-2):115-26.

[193] Abrantes J, Lopes AM, Dalton KP, Parra F, Esteves PJ. Detection of RHDVa on the Iberian Peninsula: Isolation of an RHDVa strain from a Spanish rabbitry. Arch Virol. 2014;159(2):321-6.

[194] Asgari S, Hardy JRE, Cooke BD. Sequence analysis of rabbit haemorrhagic disease virus (RHDV) in Australia: Alterations after its release. Arch Virol. 1999;144(1):135-45.

[195] Kovaliski J, Sinclair R, Mutze G, Peacock D, Strive T, Esteves PJ, et al. ( RHDV ) in Australia : when one became many. 2015;23(2):408-20.

[196] Elsworth P, Cooke BD, Kovaliski J, Sinclair R, Holmes EC, Strive T. Increased virulence of rabbit haemorrhagic disease virus associated with genetic resistance in wild Australian rabbits (*Oryctolagus cuniculus*). Virology. 2014;464-465(1):415-23.

[197] Silvério D, Lopes AM, Melo-Ferreira J, Magalhães MJ, Monterroso P, Serronha A, et al. Insights into the evolution of the new variant rabbit haemorrhagic disease virus (GI.2) and the identification of novel recombinant strains. Transbound Emerg Dis. 2018;65(4):983-92.

[198] Esteves PJ, Abrantes J, Bertagnoli S, Cavadini P, Gavier-Widén D, Guitton JS, et al. Emergence of Pathogenicity in Lagoviruses: Evolution from Pre-existing Nonpathogenic Strains or through a Species Jump? PLoS Pathog. 2015;11(11):56-61.

[199] Capucci L, Cavadini P, Schiavitto M, Lombardi G, Lavazza A. Increased pathogenicity in rabbit haemorrhagic disease virus type 2 (RHDV2). Vet Rec. 2017;180(17):426.

[200] Lopes A, Dalton KP, Magalhães M, Parra F, Esteves P, Holmes E, et al. Full genomic analysis of new variant rabbit hemorrhagic disease virus revealed multiple recombination events. J Gen Virol. 2015;96(Pt_6):1309-19.

[201] Hu B, Wang F, Fan Z, Song Y, Abrantes J, Zuo Y, et al. Recombination between G2 and G6 strains of rabbit hemorrhagic disease virus (RHDV) in China. Arch Virol. 2017;162(1):269-72.

[202] Hall RN, Mahar JE, Read AJ, Mourant R, Piper M, Huang N, et al. A strain-specific multiplex RT-PCR for Australian rabbit haemorrhagic disease viruses uncovers a new recombinant virus variant in rabbits and hares. Transbound Emerg Dis. 2018;65(2):e444-56.

[203] Abrantes J, Esteves PJ, Van Der Loo W. Evidence for recombination in the major capsid gene VP60 of the rabbit haemorrhagic disease virus (RHDV). Arch Virol. 2008;153(2):329-35.

[204] Mahar JE, Nicholson L, Eden J, Duchêne S, Kerr PJ, Duckworth J. Benign Rabbit Caliciviruses Exhibit Evolutionary Dynamics Similar to. J Virol. 2016;90(20):9317-29.

[205] Forrester NL, Moss SR, Turner SL, Schirrmeier H, Gould EA. Recombination in rabbit haemorrhagic disease virus: Possible impact on evolution and epidemiology. Virology. 2008;376(2):390-6.

[206] Camarda A, Pugliese N, Cavadini P, Circella E, Capucci L, Caroli A, et al. Detection of the new emerging rabbit haemorrhagic disease type 2 virus (RHDV2) in Sicily from rabbit (*Oryctolagus cuniculus*) and Italian hare (*Lepus corsicanus*). Res Vet Sci. 2014;97(3):642-5.

[207] Jackson RJ, Hall DF, Kerr PJ. Construction of recombinant myxoma viruses expressing foreign genes from different intergenic sites without associated attenuation. J Gen Virol. 1996;77(7):1569-75.

[208] Hall AJ, Rosenthal M, Gregoricus N, Greene S a., Ferguson J, Henao OL, et al. Incidence of acute gastroenteritis and role of norovirus, Georgia, USA, 2004-2005. Emerg Infect Dis. 2011;17(8):1381-8.

[209] Hall RN, Mahar JE, Haboury S, Stevens V, Holmes EC, Strive T. Emerging Rabbit Hemorrhagic Disease Virus 2 (RHDVb), Australia. Emerg Infect Dis. 2015 Dec;21(12):2276-8.

[210] Le Gall-Reculé G, Lemaitre E, Bertagnoli S, Hubert C, Top S, Decors A, et al. Large-scale lagovirus disease outbreaks in European brown hares (*Lepus europaeus*) in France caused by RHDV2 strains spatially shared with rabbits (*Oryctolagus cuniculus*). Vet Res. 2017;48(1):1-9.

[211] Hall R, Peacock D, Kovaliski J, Mahar J, Mourant R, Piper M, et al. Detection of RHDV2 in European brown hares (*Lepus europaeus*) in Australia. Vet Rec. 2017;180(5):121.

[212] Bell D, Davis J, Gardner M, Barlow A, Rocchi M, Gentil M, et al. Rabbit haemorrhagic disease virus type 2 in hares in England. Vet Rec. 2019;184(4):127-8.

[213] Rocchi M, Maley M, Dagleish M, Boag B. Rabbit haemorrhagic disease virus type 2 in hares in Scotland. Vet Rec. 2019;185(1):23.

[214] Buehler M, Jesse ST, Kueck H, Lange B, Koenig P, Jo WK, et al.

Lagovirus europeus GI.2 (rabbit hemorrhagic disease virus 2) infection in captive mountain hares (*Lepus timidus*) in Germany. BMC Vet Res. 2020;16(1):1-6.

[215] Neimanis AS, Ahola H, Larsson Pettersson U, Lopes AM, Abrantes J, Zohari S, et al. Overcoming species barriers: An outbreak of Lagovirus europaeus GI.2/RHDV2 in an isolated population of mountain hares (*Lepus timidus*). BMC Vet Res. 2018;14(1):1-12.

[216] Rouco C, Abrantes J, Delibes-Mateos M. Lessons from viruses that affect lagomorphs. Science (80- ). 2020;369(6502):386.

[217] Asin J, AC N, Moore J, V G-A, Clifford D, Lantz E, et al. Outbreak of rabbit hemorrhagic disease virus 2 in the southwestern United States: first detections in southern California. J Vet Diagn Invest. 2021;33(4):728-31.

[218] Lopes AM, Marques S, Silva E, Magalhães MJ, Pinheiro A, Alves PC, et al. Detection of RHDV strains in the Iberian hare (*Lepus granatensis*): Earliest evidence of rabbit lagovirus cross-species infection. Vet Res. 2014;45(1):1-7.

[219] Tan M, Hegde RS, Jiang X. The P Domain of Norovirus Capsid Protein Forms Dimer and Binds to Histo-Blood Group Antigen Receptors. J Virol. 2004;78(12):6233-42.

[220] Guillon P, Ruvoën-clouet N, Le Moullac-Vaidye B, Marchandeau S, Le Pendu J. Association between expression of the H histo-blood group antigen, α1,2fucosyltransferases polymorphism of wild rabbits, and sensitivity to rabbit hemorrhagic disease virus. Glycobiology. 2009;19(1):21-8.

[221] Ruvoën-Clouet N, Ganière JP, André-Fontaine G, Blanchard D, Le Pendu J. Binding of Rabbit Hemorrhagic Disease Virus to Antigens of the ABH Histo-Blood Group Family. J Virol. 2000;74(24):11950-4.

[222] Le Pendu J, Nyström K, Ruvoën-Clouet N. Host-pathogen co-evolution and glycan interactions. Curr Opin Virol. 2014;7(1):88-94.

[223] Leuthold MM, Dalton KP, Hansman GS. Structural analysis of a rabbit hemorrhagic disease virus binding to histo-blood group antigens. J Virol. 2015;89(4).

[224] Hammerschmidt F, Schwaiger K, Dähnert L, Vina-Rodriguez A, Höper D, Gareis M, et al. Hepatitis E virus in wild rabbits and European brown hares in Germany. Zoonoses Public Health. 2017;64(8):612-22.

[225] Wang L, Liu L, Wang L. An overview: Rabbit hepatitis E virus (HEV) and rabbit providing an animal model for HEV study. Rev Med Virol. 2018;28(1):e1961.

[226] Caballero-Gómez J, García Bocanegra I, Gómez-Guillamón F, Camacho-Sillero L, Zorrilla I, Lopez-Lopez P, et al. Absence of Hepatitis E virus circulation in wild rabbits ( *Oryctolagus cuniculus* ) and Iberian hares ( *Lepus granatensis* ) in Mediterranean ecosystems in Spain. Transbound Emerg Dis. 2020;67(4):1422-7.

[227] De Matos R, Russell D, Van Alstine W, Miller A. Spontaneous fatal Human herpesvirus 1 encephalitis in two domestic rabbits (*Oryctolagus cuniculus*). J Vet Diagnostic Investig. 2014;26(5):689-94.

[228] Hinze HC. Induction of lymphoid hyperplasia and lymphoma-like disease in rabbits by herpesvirus sylvilagus. Int J Cancer. 1971;8(3):514-22.

[229] Hinze HC. New Member of the Herpesvirus Group Isolated from Wild Cottontail Rabbits. Infect Immun. 1971;3(2):350-4.

[230] Jin L, Löhr C V., Vanarsdall AL, Baker RJ, Moerdyk-Schauwecker M, Levine C, et al. Characterization of a novel alphaherpesvirus associated with fatal infections of domestic rabbits. Virology. 2008;378(1):13-20.

[231] Abade dos Santos FA, Monteiro M, Pinto A, Carvalho CL, Peleteiro MC, Carvalho P, et al. First description of a herpesvirus infection in genus Lepus. PLoS One. 2020;15(4):1-20.

[232] Hesselton RM, Yang WC, Medveczky P, Sullivan JL. Pathogenesis of Herpesvirus sylvilagus infection in cottontail rabbits. Am J Pathol. 1988;133(3):639-47.

[233] Babra B, Watson G, Xu W, Jeffrey BM, Xu JR, Rockey DD, et al. Analysis of the genome of leporid herpesvirus 4. Virology. 2012;433(1):183-91.

[234] Wibbelt G, Frollich K. Infectious Diseases in European Brown Hare (*lepus europaeus*). Wildl Biol Pract. 2005;1(1).

[235] Martella V, Moschidou P, Pinto P, Catella C, Desario C, Larocca V, et al. Astroviruses in rabbits. Emerg Infect Dis. 2011;17(12):2287-93.

[236] Cerioli M, Lavazza A. Viral enteritis of rabbits. Recent Adv Rabbit Sci. 2006;(August):181-6.

[237] Solans L, Arnal JL, Sanz C, Benito A, Chacón G, Alzuguren O, et al. Rabbit Enteropathies on Commercial Farms in the Iberian Peninsula: Etiological Agents Identified in 2018-2019. Animals. 2019;9(12):1142.

[238] De Leener K, Rahman M, Matthijnssens J, Van Hoovels L, Goegebuer T, Van Der Donck I, et al. Human infection with a P[14], G3 lapine rotavirus. Virology. 2004;325(1):11-7.

[239] Bonica MB, Zeller M, Van Ranst M, Matthijnssens J, Heylen E. Complete genome analysis of a rabbit rotavirus causing gastroenteritis in a human infant. Viruses. 2015;7(2):844-56.

[240] Small JD, Woods RD. Relatedness of Rabbit Coronavirus to Other Coronaviruses. In: Coronaviruses. Springer US; 1987. p. 521-7.

[241] Descôteaux J-P, Lussier G, Berthiaume L, Alain R, Seguin C, Trudel M. An enteric coronavirus of the rabbit: Detection by immunoelectron microscopy and identification of structural polypeptides. Arch Virol. 1985;84(3-4):241-50.

[242] Buonocore M, Marino C, Grimaldi M, Santoro A, Firoznezhad M, Paciello O, et al. New putative animal reservoirs of SARS-CoV-2 in Italian fauna: A bioinformatic approach. Heliyon. 2020;6(11):e05430.

[243] Mykytyn AZ, Lamers MM, Okba NMA, Breugem TI, Schipper D, Van Den Doel PB, et al. Susceptibility of rabbits to SARS-CoV-2. Emerg Microbes Infect. 2021;10(1):1-7.

[244] Stenglein MD, Velazquez E, Greenacre C, Wilkes RP, Ruby J, Lankton JS, et al. Complete genome sequence of an astrovirus identified in a domestic rabbit (*Oryctolagus cuniculus*) with gastroenteritis. Virol J. 2012;9(1):216.

[245] Mahar JE, Shi M, Hall RN, Strive T, Holmes EC. Comparative Analysis of RNA Virome Composition in Rabbits and Associated Ectoparasites. J Virol. 2020;94(11).

[246] Lanave G, Martella V, Farkas SL, Marton S, Fehér E, Bodnar L, et al. Novel bocaparvoviruses in rabbits. Vet J. 2015;206(2):131-5.

Chapter 3

# Housing and Rabbit Welfare in Breeding Does

*Arantxa Villagrá García*

## Abstract

Animal welfare is a rising concern in livestock production and to assess the welfare state of an animal, it is needed to consider health, behaviour and emotions. Housing conditions and management normally impair animal welfare at different levels, so developing housing systems and management practices that imply a high level of animal welfare by preserving productive results is a need. Rabbit cages have to be improved in terms of space, enrichment and materials used to promote better conditions, and several alternatives are under evaluation such as increasing available space, providing animals with elevated platforms and hiding spaces, making available gnawing materials or changing cage materials. Moving from individual to collective housing systems to enhance social interaction is also being tested. Most of these alternatives have provided some steps towards better housing and management conditions for rabbits, while some of them have demonstrated to create more problems and have been abandoned. There is still a long way to go on this research topic.

**Keywords:** behaviour, breeding does, environmental enrichment, group housing, housing, welfare

## 1. Introduction

Consumers, supermarkets, producers, politicians and other stakeholders are increasingly using the term animal welfare regarding livestock, as the public interest for animal welfare has been arising over the last four decades, especially in Europe. Rabbit rearing must adapt to these concerns, and re-thinking and re-organising husbandry systems seem needed.

In this chapter, we will explore animal welfare in rabbit production from a scientific perspective. The basics of animal welfare concept and evaluation will be described in depth, providing a scientific basis to assess such a complex topic. Rabbits have singular behavioural patterns, physical conditions and social relationships, not comparable to other livestock species, thus leading to specific needs in terms of housing and management.

Different housing systems will be explored throughout the chapter, considering the implication on animal welfare and productivity. Both present and new approaches will be considered.



IntechOpen

## 2. Animal welfare

### 2.1 Definition and importance

Animal welfare is a term that has been created to express the ethical concerns of society about the animal's way of living, so it responds to social demands. However, it is a scientific concept [1], and there is a scientific method to identify, interpret and implement societal concerns about animal quality of life [2].

Due to this social and ethical origin, there might be as many definitions of animal welfare as people who are asked about it. In fact, it has been concluded [2] that it was impossible to give welfare a precise scientific definition. However, the most accepted definition of animal welfare is "the state of an animal as it attempts to cope with its environment, and the feelings associated to it" [3]. However, John Webster in [4] defends that "animal welfare is determined by its capacity to avoid suffering and sustain fitness", which modifies in some aspects Broom's definition, as it gathers how animals feel with the humans' responsibility. This fits with the "Five Freedoms" concept, which are the pillars of animal welfare nowadays and the base from which certifying protocols such as Welfare Quality® or AWIN® have aroused.

Five Freedoms are formulated as follows:

- Freedom from thirst, hunger and malnutrition
- Freedom from discomfort
- Freedom from pain, injuries and disease
- Freedom from fear and distress
- Freedom to express normal behaviour

These Five Freedoms refer to the ideal state, but they also might be interpreted as an absolute standard of good welfare, which prevents animal suffering [4].

### 2.2 Animal welfare in farmed animals, including rabbits

It has to be considered that there are several welfare problems that are inherent to animal production, mainly the lack of freedom of choice [4]. This lack of choice options is especially remarkable in caged systems, as behavioural repertoire is critically impaired mainly due to the lack of space or the isolation, so animal production has been deeply questioned in the last years.

In order to respond to this situation, the Council of the European Union (EU) has developed regulations for most of the farm species: Council Directive 98/58/EC for protection of animals kept for farming purposes, Council Directive 1999/74/EC (minimum standards for the protection of laying hens), Council Directive 2007/43/EC (minimum rules for the protection of chickens kept for meat production), Council Directive 2008/119/EC (minimum standards for the protection of calves) and Council Directive 2008/120/EC (minimum standards for the protection of pigs).

However, in the case of rabbits, the European Council has drafted 18 times the "Draft recommendation concerning domestic rabbit – *Oryctolagus cuniculus*" that was never definitively published. Thus, no regulation has been issued for the protection of farmed rabbits [5].

In the absence of EU directives, some member countries have drafted their own legislation or recommendations on the housing of farmed rabbits opting even for the abolition of cage housing systems.

If we want to speak about rabbit welfare, it is crucial to know some aspects about the natural behaviour of wild rabbits.

Rabbit domestication is very recent and there have not been many changes as compared to the wild rabbit, so the behaviour of domestic rabbit is still very similar to wild rabbit, and the biological needs are the same so, under commercial conditions, where specific behaviours or social relationships are impaired, there could arise a welfare problem [6]. Rabbits are social and gregarious animals, and, in wild conditions, they live in groups with more females than bucks (1–4 males and 1–9 females), except around giving birth, when they separate from the rest of the group: the dominant doe uses the common burrow and the rest of the females have to build their nest in different places [7] by mixing their own hair with grass or any other material they can find. The rabbits spend much time resting with conspecifics and in close contact, and fights are produced mainly by hierarchies [8]. However, young rabbits are easily tolerated if the group is not too big [9].

Olfaction is crucial for rabbits both in sexual and social behaviours, and hearing is also an important sense: big and mobile ears are useful for the detection of predators [9]. This characteristic of predated animal determines many behaviours of both wild and commercial rabbits. Ears also participate in body temperature regulation [10], and regarding to the vision, rabbits have a panoramic vision although not very precise [11].

On the other hand, rabbits have a unique feeding behaviour as compared to other species. Their feeding varies along the day and approximately 60% of the solid ingestion (except for caecotrophs) takes places during the dark period [12]. In addition, they also practice caecotrophy, ingesting soft faeces directly from the anus, mainly during the early morning. Finally, although rabbits are not rodents, one of their essential characteristics is the need for gnawing.

Rabbits in nature move throughout small hops (hopping), although if they are under alert, they can also walk gently. Playing behaviour is also common among kits, as well as scratching associated to the building of the hutch or exploration. Rabbits usually rest by lying down on the belly with the back legs stretched or shrink depending on if they are in alert or resting state, respectively. They can also lie laterally, which indicates a maximum level of relaxation. They usually rest between 12 and 18 hours per day and they prefer resting in groups or sustained on a firm surface [9]. Another important inherent activity is self-care of their hair (it could be considered as self-grooming) by using their legs, teeth and tongue. Moreover, the possibility of grooming other members of the group is important to its cohesion [9].

In addition, rabbits explore their area by sniffing, although this activity is frequently interrupted by stimuli that they perceive as threatening. When they get this alert alarm, they adopt specific postures such as sitting or standing on their back legs with the ears completely erect and orientated to the source of the stimulus. Another pattern related to these behaviours are freezing (total immobility of the animal) or kicking the floor or the walls to let the congeners know about a potential danger. These alert behaviours are especially important because they are predated animals, so they have a strong need to be alert and alert the conspecifics about potential damages.

As concern to social behaviour, wild rabbits live in groups and hierarchies are clearly established throughout aggressive interactions. According to [13], when kits go outside their mothers' nest area, they may come into contact with hostile conspecifics, but it is not until sexual maturity when aggressive behaviours become more important [14]. The hierarchic ranch is maintained over time within a stable group.

Regarding sexual behaviours, the evolution success of rabbits is due to "the number" [15], which means that wild rabbits tend to have big and frequent litters. In terms of maternal behaviour, attention paid to the litter is scarce, and the preparation of the nest 2–3 days before parturition and lactating once a day for a few minutes [16] are the most remarkable related activities. Once the kits have been fed, the doe hides and leaves the nest and comes back for the next suckling event. Wild breeding doe opens the nest 18–20 days post-partum to allow the kits to go out and start to eat some solid food. When the kits are 24–25 days old, the rabbit does abandon the nest to be devoted to the next litter [13].

In general, some behaviours such as gnawing, hopping, social interactions, building the nest, lactating once per day and alert behaviours are a need for domestic rabbits, so they have to be considered when housing facilities for breeding does are designed.

## 2.3 Measurement of animal welfare

When we speak about animal welfare, it is important to remind that, as explained before, there are scientific methods to assess it. In general, there are different types of indicators of animal welfare that have been traditionally used:

- Resource-based indicators, which provide information about housing itself, such as temperatures, space allowance, air quality, noise, etc.

- Management-based indicators, which are related to managing practices such as handling or feeding, but also to specific practices of each specie such as dehorning or castration

- Animal-based indicators, which are those measured directly on the animals, such as physiological indicators, lesions, lameness or behaviour

However, there is no golden standard of welfare indicators, and each approach has to be adapted, depending on the objectives of the assessment, the conditions of the farm or even the requirements of the market. Nevertheless, it is generally accepted that animal-based measurements are the most useful and provide the most valuable information and approximation to the real state of the welfare of the animals [4]. Nevertheless, this does not mean that resource- and management-based measurements are not useful anymore, but they need to be combined with other more specific measurements.

Animal welfare indicators can be divided into two big groups: indicators for short-term and long-term problems.

It has been defined [17] that short-term measures are, for example, heart rate or plasma cortisol concentration, while some measures of behaviour, immune system function and disease state are long-term measurements. Housing conditions usually affect long-term indicators, as they act as potential chronic stress source. Transport or handling would be examples of short-term problems.

Some of the measures that can be performed to assess animal welfare are productivity, maintenance behaviours, abnormal behaviours, other behaviours such as maternal interactions, endocrine measures of stress, blood pressure, heart rate and respiratory rate, incidence of disease, level of immune protection and bone strength, and rate of injury and wounding [18]. As it is seen, all of them are animal-based measurements and display a multidisciplinary approach to animal welfare evaluation.

However, when housing conditions are evaluated, there is one aspect that has to be especially considered: the assessment of feelings, which means to know the real importance of the studied aspect for the animal. The main way to assess feelings in animals is through preference tests, in which the animals are allowed to choose between different possibilities. It is assumed that animals will make their choices according to how they feel [2] and we consider that animals make choices that are in

**Figure 1.**
*Motivation cage constructed to assess the preference of breeding does for different sizes of cages.*

their own best interests [19]. However, these tests can be influenced by the specific conditions in which they are performed, such as temperature, age, time of the day, season or previous experience. There is a tool to avoid these effects, and it is through motivation tests in which the strength of preference is assessed, and the animal gets a reward in response to some work [19]. An example of this test is shown in **Figure 1**, in which the rabbits could choose between three different housing cages, and ballasted push doors give access to each cage. Another possibility is the reactivity tests, in which the reaction or the fear to an environment or other stimulus is assessed (e.g., open field tests or tonic immobility tests [19]).

In summary, to develop a proper animal welfare assessment, physiological, behavioural and emotional needs have to be taken into account.

## 3. Conventional housing of breeding does

### 3.1 Development of current systems

Taking into account the behavioural repertoire of rabbits recently explained, one of the major concerns of commercial rabbits' welfare is related to housing and their equipment, as the restriction of the space may impair severely the development of specific natural behaviours, reduce the level of activity or increase the appearance of abnormal behaviours [20]. In the present commercial productive system, breeding rabbits (both does and bucks) are commonly housed in individual barren wire mesh cages, with no bedding material, although in alternative or organic systems, some straw-bedded pens can be found and does have a compartment to give birth (nest) and share the cage with the kits once they leave this nest, and they do not have any possibility to avoid each other during this period. These types of wire systems were developed in the 1960s due to hygienic reasons, as they allowed the separation between the animals and their faeces and urine. Nevertheless, these types of housing systems are being deeply criticised mainly in the European Union, and their evolution is being questioned. This evolution can be both to systems in which rabbits are housed alone or group housing systems, as we will see further.

The main welfare problem of individual cages is related to the impairment of certain natural behaviours, while the rest of the Five Freedoms are obviously guaranteed. However, some specific behaviours are allowed in caged systems, mainly all those related to peripartum and lactation, as a nest is allocated in the cage 2–3 days prior to the parturition to allow the doe to prepare the nest by mixing her hair with any provided material such as cotton, straw or wood shavings. Nevertheless, it seems that does have their preferences according to these materials, and when they can choose, they prefer straw to build the nest as compared to other materials such as wood shavings [21]. This could be due to the similarity between straw and the material they use in wild conditions.

In fact, the main concerns about these cages are the isolation of the rabbits and especially, the dimensions of the cages. Dimensions of individual cages vary between countries according to **Table 1** [8].

Some problems related to these cages are excess of lying time, locomotive problems or abnormal skeleton development [8]. But not only the width and the length are a possible problem, but also cage height, as it can avoid the development of alert behaviours such as sitting or standing with the ears erect.

However, although cages are the main concern in rabbit welfare nowadays, there are other housing conditions that may negatively affect the welfare of the animals. In fact, the environment that surrounds the animals and their characteristics are critical for animal welfare, and inadequate environmental conditions can favour the appearance of stress or sanitary problems. There are then five key aspects to control in the rabbits' facilities: temperature, relative humidity, air velocity, concentrations of dust and gases and lighting.

### 3.2 Foot mats

Conventionally, as it was said, rabbits were housed in wire mesh cages. In this type of floor, the incidence of pododermatitis (sore hocks) is high (up to 71.5% of animals [22] and 86.7% [23]), mainly due to does' weight and the long time spent in the same cage. Pododermatitis is a skin illness in rabbits that mainly appears in the back legs and causes pain and suffering to farmed rabbits [9]. Once it appears,

| Country/Type of cage | Width (cm) | Length (cm) | Height (cm) | Available surface ($cm^2$) |
|---|---|---|---|---|
| **France** | | | | |
| Young females | 26–30 | 45–50 | 29–30 | 1200–1500 |
| Lactating female with litter | 40 | 90–100 | 29–30 | 3600–4000 |
| **Italy/Hungary** | | | | |
| Young females | 38 | 43 | 35 | 1600 |
| Lactating female with litter | 38 | 95 | 35 | 3600 |
| **Spain** | | | | |
| Young females | 30 | 40 | 33 | 1200 |
| Lactating female with litter | 40 | 85 | 33 | 3400 |
| **EFSA recommendations (2005)** | | | | |
| Breeding males and females | 38 | 65–75 | 38–40 | 3600 |

*NOTE: Dimensions without nest.*

**Table 1.**
*Summary of the dimensions of individual cages for breeding does in the main rabbit meat producers, European countries.*

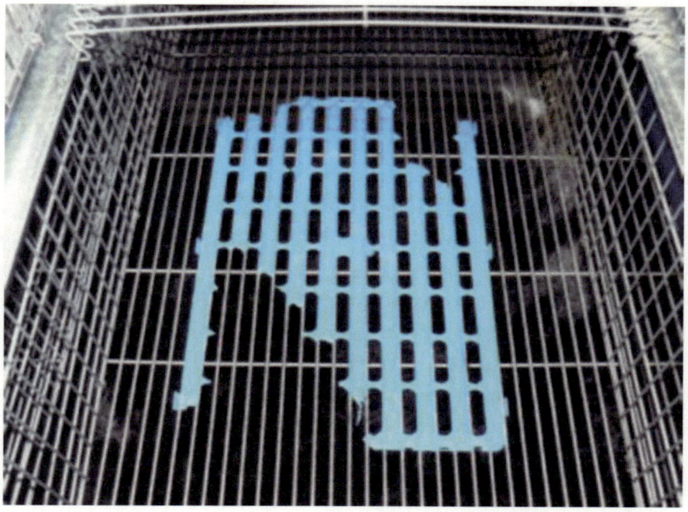

**Figure 2.**
*Gnawed foot mat.*

the wound can be colonised by pathogen microorganisms, enhancing the severity of this welfare problem. In fact, according to [9], it is considered that 16.5% of the culling rate in rabbit farms is due to pododermatitis.

In a deep study carried out by [24] about foot mats, they showed a positive relationship between their use and animal welfare. They found a significant reduction (81.3%) in the prevalence of sore hocks in farms where foot mats were used. This could indicate that the rabbits are more comfortable on this type of flooring and this seems to be confirmed by other studies [25] that found that rabbits clearly preferred foot mats over wire mesh when they can choose. By all these reasons, they became a "must" in the commercial rabbit houses nowadays, but it is important that they are in a good state of conservation, and foot mats gnawed or in bad conditions as that shown in **Figure 2** have to be avoided.

## 4. New models/trends in housing systems for breeding does

### 4.1 Individual systems with improved environments

Most of the studies that continue working on individual systems for breeding does focus now on environmental enrichment.

Environmental enrichment consists of adding complexity to the environment in which the animals live, by providing different elements that can help the development of certain behaviours, which cannot be performed without those additional elements. Thus, environmental enrichment can improve the quality of life of animals in captivity, as it allows them to fulfil specie-specific behaviours. For rabbits, environmental enrichment can favour activities such as gnawing, scratching, hopping and hiding. The main elements used to enrich rabbits' environment are as follows:

- Bigger cages: some standard cages have been modified to give the animals more available space to move freely. These cages can or cannot have more additional elements. Does in larger spaces are more active, as they perform more active behaviours such as locomotive ones [26], and when they are allowed to choose between different sizes of cages, they seem to be motivated for longer and higher cages [27]. The importance of height can be due especially to the possibility of performing alert and exploratory behaviours.

- Gnawing elements: wood sticks or synthetic materials hanging on the walls of the cages or on the floor. These elements satisfy one of the most important needs of the rabbits, and it is performed independently of the housing system in which the animals are bred (see **Figure 3**). They also help to avoid the excessive growing of incisors [28], which is helpful especially in group systems, in order to decrease the intensity of the injuries when aggressive episodes take place. Moreover, this type of enrichment also leads to hygienic problems, especially when wooden sticks are used. As a consequence, there is a recommendation to fix the sticks on the walls or the ceiling of the cage instead of the floor, in order to avoid an undesirable contamination of the sticks by pathogenic microorganisms that could cause the animals severe illnesses [29].

- Refugees/hiding places: this type of enrichment is especially important when the breeding does are housed in groups, as they provide a place to the animals when they are threatened, so they can reduce the number of aggressions [30]. In fact, hiding places can even help to reduce the number of does culled as a consequence of aggressive interactions [31].

- Mirrors: they are a source of sensory enrichment, which is especially important for rodents and rabbits [32]. According to recent studies [33], mirrors act

**Figure 3.**
*Gnawing sticks previous to allocation in a cage and after 1 month of use.*

positively on the behavioural repertoire of the rabbits, as they might reduce the effects of isolation and they compensate the lack of social contact [34], so they become an interesting enrichment device in individual housing systems, although group living rabbits also showed preference for cages half covered with mirrors [35].

- Platforms (**Figure 4**): this is the most-studied enrichment in rabbit housing systems. According to different studies, does prefer jumping on elevated places if they exist [36, 25] and they use them to escape from the kits once they have left the nest.

According to our own unpublished results, breeding does use the platform mainly during the first 2 weeks of lactation, when the kits are still in the nest. During this period, the percentage of time spent on the platform reaches to 15% of time (**Figure 5**). During the third week of lactation, kits start to leave the nest, but they still cannot go up to the platform, so the doe uses it almost 35% of time, presumably to avoid the kits [37]. On the contrary, from the fourth week of lactation, the kits can rise to the platform and the percentage of time they spend on it is up to 66%, whereas the time of the doe decreases to 7% of time. From this moment, the use of the platform by the doe decreases and the kits continue using it up to 94% of the time, being clear the exclusion effect of the platform between does and kits. Moreover, some problems derived from the use of an elevated platform have been observed, mainly related to hygiene, as they can defecate and urinate in the platform (and beneath) and thus, the level of cleaning of both does and kits is reduced and the possibility of infection rises, as the animals are in contact with their faecal material [25, 38]. Daily health checking is also impaired because the animals are less visible and handling of the animals becomes more dangerous as does are hidden below the platform and defensive attitudes can be developed.

**Figure 4.**
*Cages with elevated platforms in which breeding does standing and with erect ears can be observed.*

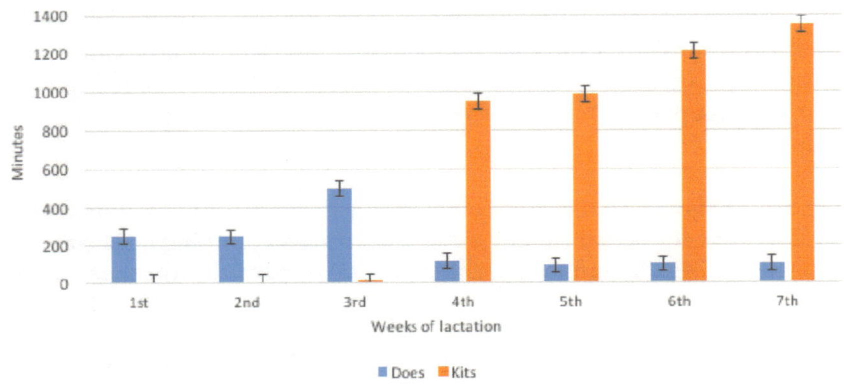


**Figure 5.**
*Time spent by the breeding does and the kits on the elevated platform as the lactation period advances (unpublished results).*

- Plastic floor, plastic mesh or slat: wire floor cages are being banned in certain Northern European countries such as Belgium or Switzerland [39], so there are several studies using alternative materials, such as slatted plastic floor. However, not all of them are suitable for countries such as Germany or Austria, where legislation requires specific dimensions of gaps. In general, common slats present good results both in production and leg health, but for example, slats with circular holes are not suitable for rabbit husbandry due to the level of dirtiness [39]. In general, the presence of plastic-slatted floors also decreases sore hock problems and improves doe welfare conditions [24].

- Litter flooring: some authors have studied the impact of housing rabbits on litter, but results show an impairment of productivity as well as enteric disorders. Ulcerative pododermatitis can also be increased when the rabbits are housed on litter, mainly in the hind legs [5]. In addition, choice tests do not show a preference for straw flooring as compared to wire and plastic flooring [40].

- Other sensory enrichment such as music has been revealed as useful to improve the well-being of the animals when they are housed in cages [41].

NOTE: some of these enrichment aspects have been implemented in the "WRSA cages", which are being used in newly renovated farms along Europe [5].

### 4.2 Collective housing

During the last years, there have been some approaches to develop group housing systems for breeding does. This grouping approach relies on the fact that wild does feed their kits once per day and they spend the rest of the day sharing time with conspecifics, leaving the kits in the nest.

When continuous group housing systems have been evaluated, several problems have been found, leading to unacceptable results in terms of welfare and productivity. Some of these were infertility, pseudopregnancy, high kit mortality and aggression [5], difficult health control, behaviour abnormalities, replacement of the does, higher productive costs [36], shorter lifespan and higher culling rate [5, 8]. Moreover, relationships among does are difficult in the first days after parity, and kit mortality is very high because of competition among does. They compete for the nests, attack the kits and sometimes raising of 2 or 3 does in the same nest box is found [5]. This can be related to the previous explained theory about "the number", as rabbit does try to guarantee the success of their own litter, even if they have to attack other does' litters.

So the main reasons for the failure of this housing system are the very high rates of aggression among females and injured does and kits.

As a consequence, some Belgian research groups in 2011 (mainly ILVO and Ghent University, [42]) examined the possibility of a part-time group housing system, in which the does are kept in groups only some part of the lactation period (**Figure 6**). The does are group housed while they are pregnant, and 2 or 3 days before parity, they are separated (normally by closing a removable wall in their home park). They give birth and live with their kits during part of the lactation period, and between 11 and 18 days, the walls are again removed and the does are mixed. This means that they cannot fight at the peripartum and they cannot hurt each others' litters [43].

In this type of systems, does spend more time moving, sniffing and grooming, mainly after grouping [44]. However, aggressions are still present although the does give birth separately. Aggressions between does mainly take place in the first days after grouping, when the hierarchy has to be re-established [43, 45]. Furthermore, these fights can lead to severe injuries in the skin of the rabbits, they impair the body condition and sometimes, they have to be separated [46].

There are several factors that can affect the level of aggressiveness during the mixing process. One of them is the group size, as aggression level rises as the group size increases [47]. The other one is the age of the kits when the does are grouped. Interactions between does when the kits are 18 days old are lower than when they are 11 days old, as their capacity of moving and leaving the nest weaken the interactions between the mothers [46].

However, there are still some problems that need to be solved in these systems, such as the introduction of does to a previously formed group or the enrichment needed to allow alert and hiding behaviours. The does are commonly mixed after sexual maturity (mainly if it is necessary to remake the group), when the level of individual aggressiveness is higher [14]. Tunnels and hiding structures have been used [31], but they must ensure the possibility of inspection of the animals, as there

**Figure 6.**
*Semi-group housing system in which removable walls can be observed, as well as hiding places and straw dispensers.*

can be specific animals that hide in the structure and their level of fearfulness is so high that they do not move from that place, even dying from starvation. The revision of the animals in this type of systems is crucial, as well as the evaluation of the animals in order to separate them as soon as any especially aggressive animal is detected.

To correct some of these problems, some alternatives have been developed [48]. They proposed a system in which the breeding does are individually identified with an electronic chip, which allows the doe to enter only in its own nest (similar to the pregnant sows' system), avoiding kits' thefts and cannibalism. However, this system is valid for research purposes, but it appears unaffordable for commercial production, so new alternatives and improvements have to be searched.

## 5. Conclusions

Rabbits' welfare must be assessed from a multifactor perspective, considering productivity, health, behaviour and emotions. Current housing systems present failures when considering animal welfare. A deep research work is being developed to create new housing systems to promote an enhanced animal welfare level. Different strategies are being considered, from cage sizing, environmental enrichment, social interactions, etc. Although a golden standard has not been yet obtained, significant milestones have been achieved, which may encourage researches to keep working in this area.

## Acknowledgements

I thank my colleagues Dr. Zsolt Szendro, Dr. Angela Trocino, Dr. Stephen Hoy, Dr. Gerolamo Xiccato, Dr. Luc Maertens and Dr. Cristina Zomeño for their contribution to this chapter as a consequence to our recent collaborations in some reviews and projects' preparation.

## Author details

Arantxa Villagrá García
Animal Technology Centre, Valencian Institute of Agricultural Research
CITA-IVIA, Castellón, Spain

*Address all correspondence to: villagra_ara@gva.es

## References

[1] Broom DM. A history of animal welfare science. Acta Biotheoretica. 2011;**59**:121-137

[2] Duncan IJH. Science-based assessment of animal welfare: Farm animals. Scientific and Technical Review. 2005;**24**(2):483-492

[3] Fraser AF, Broom DM. Farm Animal Behavior and Welfare, 3rd ed. Wallingford: CAB International; 1990. p. 437

[4] Webster J. Animal Welfare: Limping Towards Eden, 1st ed. Oxford, UK: Blackwell Publishing Ltd; 2005. p. 284

[5] Zs S, Trocino A, St H, Xiccato G, Villagrá A, Maertens L. A review of recent research outcomes on the housing of farmed domestic rabbits: Reproducing does. World Rabbit Science. 2019;**27**:1-14

[6] Gunn D, Morton DB. Inventory of the behaviour of New Zealand white rabbits in laboratory cages. Applied Animal Behaviour Science. 1995;**45**(3-4):277-292

[7] González-Redondo P, González-Mariscal G, López M, Fernández-Carmona J, Finzi A, Villagrá A. Comportamiento materno y bienestar de la coneja doméstica y silvestre y su camada. ITEA. 2015;**111**(4):326-347

[8] Trocino A, Xiccato G. Animal welfare in reared rabbits: A review with emphasis on housing systems. World Rabbit Science. 2006;**14**:77-93

[9] Opinion of the scientific panel on animal health and welfare (AHAW) on a request form the Commission related to "The impact of current housing and husbandry systems on the health and welfare of farmed domestic rabbits". doi: 10.2903/j.efsa.2005.267

[10] Harkness JE, Turner PV, VandeWounde S, Wheeler CL. Harkness and Wagner's Biology and Medicine of Rabbits and Rodents. 5th ed. Iowa: Blackwell Publishing; 2010. pp. 23-45

[11] Lebas F. Biologie du lapin. 2002. Available from: http://cuniculture.info/Dcos/Biologie/biologie-09.htm

[12] Gidenne T, García J, Lebas F, Licois D. Nutrition and feeding strategy: interactions with pathology. In: De Blas C, Wiseman J, editors. Nutrition of the Rabbit. Wallingford: CAB International; 2010. pp. 179-199

[13] Dudzinski ML, Mykytowycz R, Gambale S. Behavioural characteristics of adolescence in young captive European rabbits, Oryctolagus cuniculus. Aggressive Behaviour. 1977;**3**:313-330

[14] Olivas I, Simarro L, Villagrá A. Development and establishment of an individual aggressiveness test protocol in breeding does. World Rabbit Science. 2016;**24**:321-326

[15] Pascual JJ. The role of body condition on new feeding and breeding programmes for reproductive rabbit does. In: 22nd Hungarian Conference on Rabbit Production; 26th May 2010; Kaposvar, Hungary; 2010. pp. 11-22

[16] Fernández-Carmona J, Solar A, Pascual JJ, Blas E, Cervera C. The behavior of farm rabbits around parturition and during lactation. World Rabbit Science. 2005;**13**:253-277

[17] Broom D, Fraser F. Domestic Animal Behaviour and Welfare. 4th ed. Wallingford: CAB International; 2007. p. 438

[18] McGlone JJ. Farm animal welfare in the context of other society issues: Toward sustainable systems. Livestock Production Science. 2001;**72**:75-81

[19] Fraser D, Matthews LR. Preference and motivation testing. In: Appleby MC, Hughes BO, editors. Animal Welfare. 1st ed. New York: CAB International; 1997. pp. 159-173

[20] Hansen LT, Berthelsen H. The effect of environmental enrichment on the behavior of caged rabbits (Oryctolagus cuniculus). Applied Animal Behaviour Science. 2000;**68**(2):163-178

[21] Blumetto O, Olivas I, Torres AG, Villagrá A. Use of straw and wood shavings as nest material in primiparous does. World Rabbit Science. 2010;**18**:237-242

[22] Rosell JM, de la Fuente LF. Culling and mortality in breeding rabbits. Preventive Veterinary Medicine. 2009;**88**:120-127

[23] Rommers JM, De Jong I. Technical note: Plastic mats prevent footpad injuries in rabbit doe. World Rabbit Science. 2011;**19**:233-237

[24] Rosell JM, de la Fuente LF. Effect of footrests on the incidence of ulcerative pododermatitis in domestic rabbit does. Animal Welfare. 2009;**18**(2):199-204

[25] Alfonso-Carrillo C, Martín E, de Blas C, Ibáñez MA, García-Rebollar P, García-Ruiz AI. Effect of cage type on the behavioural pattern of rabbit does ar different physiological states. World Rabbit Science. 2014;**22**:59-69

[26] Bignon L, Bouchier M, Coutelet G, Galliot P, Souchet C, Fortun-Lamothe L. Individual housing of young does in different sized cages: impact on welfare, economic costs and productive data. In: Proceedings of the 10th World Rabbit Congress; 3-6 September 2012; Sharm El-Sheikh, Egypt; 2012. pp. 1045-1049

[27] Villagrá A, Martinez-Paredes E, Martínez-Talaván A, Estellés F, Cervera C. Are breeding rabbits motivated for bigger cages? In: Proceedings of the 53rd Congress of the ISAE; 5-9 August 2019; Bergen, Norway; 2019. p.190

[28] Princz Z, Orova Z, Nagy I, Jordan D, Stuhec I, Luzi F, et al. Application of gnawing sticks in rabbit housing. World Rabbit Science. 2007;**15**:29-36

[29] Marín C, Simarro-Catalá L, Villagrá A. Technical note: Assessment of best location of gnawing sticks in growing rabbit cages. World Rabbit Science. 2018;**26**(3):249-254

[30] Baumans V, Van Loo PLP. How to improve housing conditions of laboratory animals: The possibilities of environmental refinement. Veterinary Journal. 2013;**195**:24-32

[31] Rommers J, Reuvekamp BJF, Gunnink H, de Jong IC. Effect of hiding places, straw and territory on aggression in group-housed rabbit does. Applied Animal Behaviour Science. 2014;**157**:117-126

[32] Baumans V. Environmental enrichment for laboratory rodents and rabbits: Requirement of rodents, rabbits and research. ILAR Journal. 2006;**46**(2):162-170

[33] Mastellone V, Bovera F, Musco N, Panettieri V, Piccolo G, Scandurra A, et al. Mirrors improve rabbit natural behavior in a free-range breeding system. Animals. 2019;**9**:533

[34] Edgar JL, Seaman SC. The effect of mirrors on the behavior of singly housed male and female laboratory rabbits. Animal Welfare. 2010;**19**:461-471

[35] Dalle Zotte A, Princz A, Matics Z, Gerencser Z, Metzger S, Szendro Z. Rabbit preference for cages and pens with or without mirrors. Applied Animal Behaviour Science. 2009;**116**:273-278

[36] Hoy S, Verga M. Welfare indicators. In: Maertens L, Coudert P, editors.

Recent Advances in Rabbit Sciences. 1st ed. Melle, Belgium: Institute for Agriculture and Fisheries Research; 2006. pp. 71-74

[37] Mirabito L. Logement et bien-etre du lapin: Les nouveaux enjeux. In: Proceedings of the 10 Journées de la Recherche Cunicole; Paris, France; 2003. pp. 163-172

[38] Cervera C, Gómez EA, Pérez-Fuentes S, Villagrá A. Bienestar y salud en conejas reproductoras. In: Proceedings of the 43rd Symposium de Cunicultura; 30-31st May 2018; Calamocha, Spain; 2018. pp. 33-42

[39] Tillmann K, Windschnurer I, Gamper J, Hinney B, Rülicke T, Podesser BK, et al. Welfare assessment in rabbits raised for meat and laboratory purposes in enclosures with two floor types: Perforated plastic with holes versus slats. Research in Veterinary Science. 2019;**122**:200-209

[40] Turner PV, Buijs S, Rommers J, Tessier M. Code of practice for the care and handling of rabbits: Review of scientific research on priority issues. National Farm Animal Care Council. https://edepot.wur.nl/446238. p. 69

[41] Peveler JL, Hickman DL. Effects of music enrichment on individually housed male New Zealand white rabbits. Journal of the American Association for Laboratory Animal Science. 2018;**57**(6):695-697

[42] Maertens L, Rommers J, Jacquet M. Le logement des lapins en parcs, une alternative pour les cages classiques dans un système "duo"? In: Proceedings of the 14èmes Journées de la Recherche Cunicole; 22-23 November 2011; Le Mans, France; 2011. pp. 85-88

[43] Buijs S, Maertens L, Hermans K, Vangeyte J, Tuyttens FAM. Behaviour, wounds, weight loss and adrenal weight of rabbit does as affected by semi-group housing. Applied Animal Behaviour Science. 2015;**172**:44-51

[44] Maertens L, Buijs S. Impact of housing system (cage vs. part-time housing) and floor type on rabbit doe welfare. In: Proceedings of the 11th World Rabbit Congress; 15-18 June 2016; Qingdao, China; 2016. pp. 707-710

[45] Rommers J, de Greef K. Towards part-time group-housing of lactating rabbit does? In: Proceedings of the 20th International Symposium on housing and diseases of rabbits, furproviding animals and pet animals. 17-18 May 2017; Celle, Germany; 2017. pp. 3-13

[46] Cervera C, Ródenas L, Martínez-Paredes E.Bienestar y salud de conejas y gazapos en sistemas de alojamiento individual o colectivo en semigrupo. In: Proceedings of the 43rd Symposium de Cunicultura; 30-31st May 2018; Calamocha, Spain; 2018. pp. 114-119

[47] Zomeño C, Birolo M, Zufellato A, Xiccato G, Trocino A. Aggressiveness in group-housed rabbit does: Influence of group size and pen characteristics. Applied Animal Behaviour Science. 2017;**194**:79-85

[48] Ruis M. Group housing of breeding does. In: Maertens L, Coudert P, editors. Recent Advances in Rabbit Sciences. 1st ed. Melle, Belgium: Institute for Agriculture and Fisheries Research; 2006. pp. 99-105

Chapter 4

# Effect of the Microclimatic Temperature-Humidity Index (THI) on the Productivity Performance of Rabbit

*Dennis Osarenren Asemota and Arierhire Michael Orheruata*

## Abstract

Temperature-humidity index (THI) is a useful and easy way to assess the risk to heat stress. This index combines effects of environmental temperature and relative humidity. This study was conducted to determine the temperature-humidity index and to assess its effects on breeding does' productivity potential at the rabbit house of the University of Benin Teaching and Research Farm. Data on microclimatic factors of ambient temperature (TEMP) and relative humidity (RH) of the rabbit house were obtained on monthly bases for 1 year. The year was divided into four seasons: late dry (January to March), early rain (April to June), late rain (July to September), and early dry (October to December). Temperature-humidity index was calculated using the ambient temperature and relative humidity values obtained on monthly and seasonal bases. The estimated values for TEMP were 34.78 ± 0.17, 33.41 ± 0.22, 31.17 ± 0.22, and 33.55 ± 0.19°C for late dry, early rain, late rain, and early dry seasons, respectively ($p < 0.05$), and the corresponding RH values were 56.89 ± 0.17, 57.23 ± 0.16, 59.03 ± 0.25, and 57.17 ± 0.20% ($p < 0.05$). The highest and lowest THI values were reported in late dry (32.03 ± 0.14) and in late rain (29.03 ± 0.19) ($p < 0.05$), in agreement with the higher and lower thermal stress in these seasons. In relation to the productivity of does, percentage conception rate ranged from 0 to 20% in late dry vs. from 52 to 56% in late rain. In conclusion, the most favorable season was late rain with the lowest THI and the highest productivity potential parameters.

**Keywords:** temperature-humidity index (THI), ambient temperature, relative humidity, heat stress, reproductive traits

## 1. Introduction

Microenvironment is the immediate physical environment surrounding an animal; that is the environment in the primary enclosure such as the cage, pen, or stall [1]. In recent times there seems to be global increase temperature in the tropics even to the consciousness of the livestock farmer. Environmental temperature and relative humidity can affect livestock husbandry by imposing thermal stress on them. Heat stress may be defined as any combination of environmental variables that give rise to conditions that are higher than those of the temperature range of the animal's thermal neutral zone [2]. Igono et al. [3] proposed that the temperature-humidity index (THI) could be used to evaluate the level of heat stress imposed by the environment. In order to estimate the severity of heat stress, the temperature-humidity index was proposed using both ambient temperature and relative humidity [4, 5].

Heat stress has been reported to evoke a series of drastic changes in the biological functions of rabbits which lead to impairment in production and reproduction [6–8] especially if the onset of heat is sudden [9]. Lebas et al. [10] reported that the body temperature of the domestic rabbit ranges from 18 to 26°C, with small variation of ±5°C. The acceptable range of relative humidity is considered to be 30–70% for most mammalian species [10, 11]. Basically, the thermoneutral zone reflects the range of ambient temperature at which internal temperature regulation is solely achieved by control of dry heat loss, which means that the metabolic rate is relatively constant without regulatory changes in heat production [12]. Fayez et al. [13] had summary that temperature of 21°C is the comfort zone in rabbits. Choudhary et al. [14] described the highly influential effect of the season on gestation period, kindling interval, and litter weight at weaning. Oguike and Okocha [15] reported a decrease in conception rate in does with increase in the re-mating intervals they were subjected to.

Information on the microenvironment of our livestock especially those reared intensively will help to redirect our management to ensure they are conducive. This study was therefore undertaken to determine the average temperature, relative humidity, and estimate temperature-humidity index in each month and the four seasons of experimental year under tropical-humid condition and to analyze the effects of the THI on productivity performance of does.

## 2. Material and methods

### 2.1 Location

The experiment was conducted at the Teaching and Research Farm of the University of Benin, Edo State, Nigeria. The University of Benin is located on latitude 6.02° N and longitude 5.06° E in the Humid Rain Forest Zone of Southern Nigeria, with an annual temperature range between 24.5 and 32.7°C, with a mean of 28.6°C. Annual rainfall ranges from 1498 to 3574 mm with a mean of 2430 mm. The relative humidity and daily sunshine are between 63.3 and 81.7% and 5.85 and 7.50 hours with means of 73.5% and 6.68 hours, respectively [16].

### 2.2 Housing

The orientation of the rabbit house was in east-west direction, with rabbits individually housed in a four-compartment hutches with dimensions of 60 cm high, 90 cm long, and 60 cm wide [17]. The hutches were made of wood with the floor and sides covered with wire netting. The hutches were about 90 cm above the ground.

### 2.3 Experimental animals, housing, and management practices

A total of 60 rabbits (50 does and 10 bucks) were used for the study. The experimental rabbits were housed individually in hutches made of wood and wire mesh, as described in the previous section. The hutches were located inside the rabbit unit. Each hutch has a feed and water trough made of weighted earthenware for concentrates and water, respectively. The rabbits were fed with commercial grower's mash of 17% CP and ME of 2800 kcal/kg and forage [18]. Feed and water were supplied ad libitum throughout the experimental period. They have a body weight of 1.7–1.8 kg at first mating. Semireproductive rate system of 10–14 days interval immediately after kindling date from one complete gestation to another [10] was adopted while the project lasted.

Nesting boxes were introduced at 25 days after successful mating. After kindling, the nest box was checked. Weaning was done at 4 weeks of age. At weaning, the does were taken away to another hutch. Cage identification method was used throughout the project.

### 2.4 Data collection

#### *2.4.1 Microenvironment data*

The ambient daily temperature and relative humidity were taken twice daily at 9.00 am and 2.00 pm using a mercury-in-glass thermometer and a wet and dry bulb thermometer, respectively. The data were grouped into months and seasons. The seasons were late dry (January to March), early rain (April to June), late rain (July to September), and early dry (October to December).

The rabbit house daily ambient temperature was measured in degree Celsius (°C), and the standard relative humidity in percentage (%) was obtained from a wet and dry thermometer humidity table. The temperature-humidity Index was computed using the procedure of Marai et al. [19] depicted as:

$$THI = t - \left[\left(0.31 - 0.31\left(\frac{RH}{100}\right)\right)(t - 14.4)\right] \tag{1}$$

where t is the dry bulb temperature in degrees Celsius (°C) and RH is the relative humidity in percentage/100. THI values were classified as <27.8, the absence of heat stress; 27.8–28.9, moderate heat stress; 29.0–0.0, severe heat stress; and >30.0, very severe heat stress.

#### *2.4.2 Productivity data*

The following traits were recorded:

a. Conception rate (CR, %), estimated as $\frac{\text{number of parities}}{\text{number of mating}} x\ 100$.

b. Kindling interval (KI, days), estimated as days between two consecutives parities

c. Successive mated doe (SMD)

d. Group of kits that reached sexual maturity at 22 weeks (GK)

Other doe productivity indices under a small holder system were calculated using the method of Odubote et al. [20]

e. Weaned/doe/year, estimated as LTSW × (365/KI)

f. Slaughtered/doe/year, estimated as weaned/doe/year × 0.85

g. Live weight/doe/year, estimated as slaughtered/doe/year × 1.85 kg

h. Dressed weight/doe/year, estimated as live weight/doe/year × 0.56

where LTSW = litter size at weaning; 0.85 = post-weaning survival rate; 0.56 = dressing out percentage for rabbit.

### 2.5 Statistical analysis

Temperature, relative humidity, and THI were analyzed with a model that included seasons as fixed effect (late dry, early rain, late rain, and early dry). Also, a polynomial line was fitted for temperature, relative humidity, and THI with months.

## 3. Results and discussion

The temperature (TEMP, °C), standard relative humidity (RH, %), and temperature-humidity index per month at the rabbit house of the University of Benin Teaching and Research Farm are presented in **Figures 1–3**, respectively. The prediction equation (polynomial) that gave the simplest and best fits was also indicated in the charts. The ambient temperature for the year ranged between 31 and 36°C. This study revealed that the least values in the seasons throughout the experimental were beyond the rabbit thermo-comfort zone of 20–21°C [21]. The microenvironment relative humidity was below 60%, ranging approximately from +1 to +5 up the sensitive limit of 55% for rabbit as reported by Lebas et al. [10], although the values of this study within HR range would be considered as appropriate (30–70%) for most mammalian species [11]. The months where the ambient temperature values were higher than the best-fit (polynomial) line (March, April, May, November

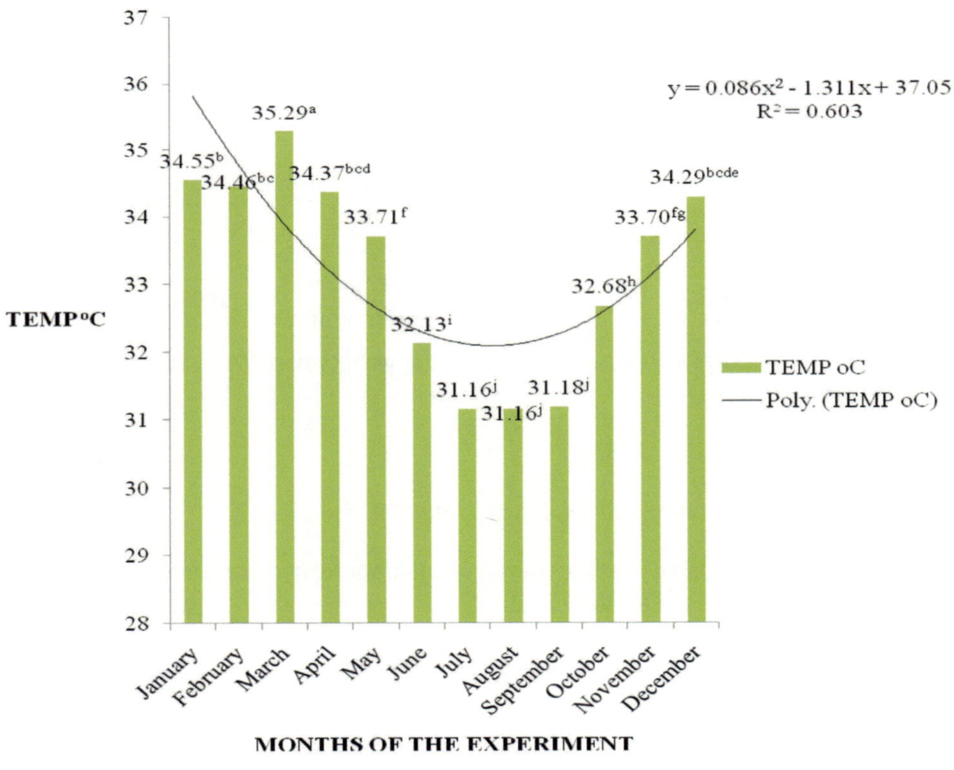


**Figure 1.**
*Temperature (TEMP, °C) during the months of the experimental periods. Mean label values are least square means, n = 12, SEM = ±0.33.* [abcdefghij]*Means with different superscripts within the chart differ significantly.*

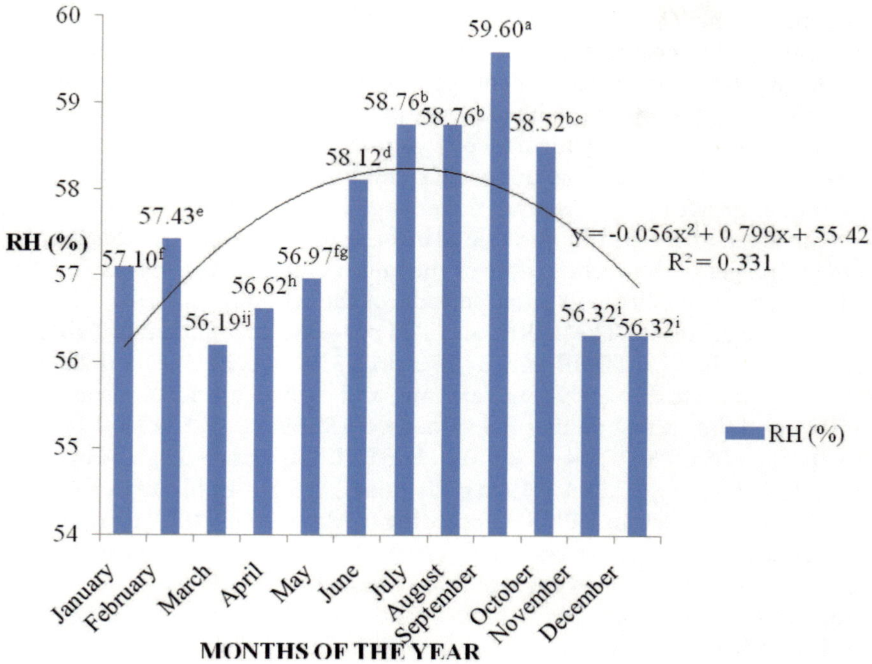


**Figure 2.**
*Relative humidity (RH, %) during the months of the experimental periods. Mean label values are least square means, n = 12, SEM = ±0.29. [abcdefghij]Means with different superscripts within the chart differ significantly.*

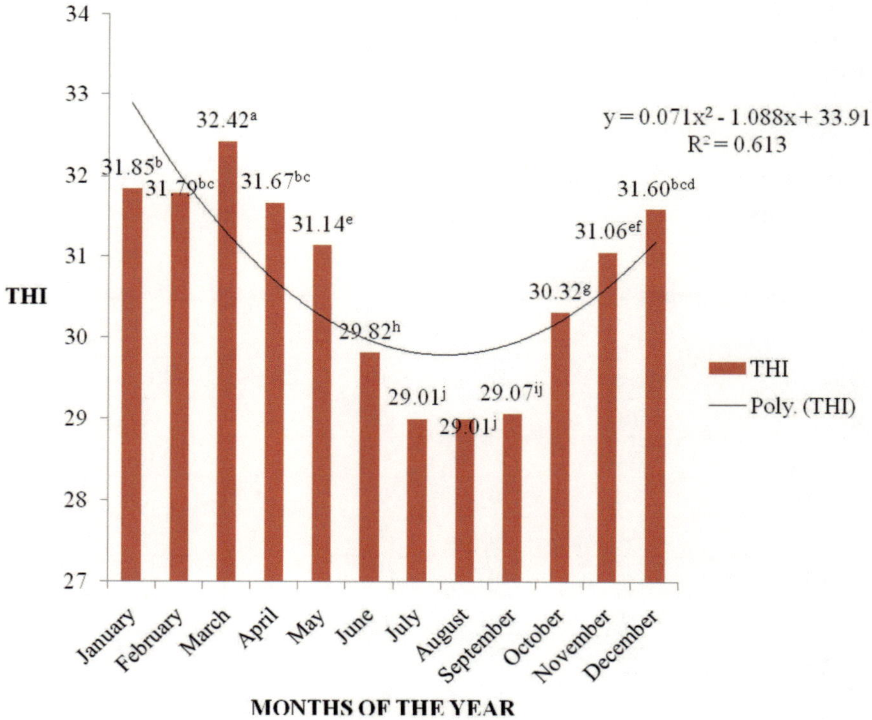


**Figure 3.**
*Temperature-humidity index (THI) during the months of the experimental periods. Mean label values are least square means, n = 12, SEM = ±0.28. [abcdefghij]Means with different superscripts within the chart differ significantly.*

and December) incidentally had lower relative humidity values. Such situation can no doubt lead to heat stress. Heat stress has been reported to cause negative balance between the net amount of energy flow from an animal to its surrounding environment and the amount of heat energy produced by the animal. Therefore, such month will affect the performance of animals. Similar observation was made by Farooq et al. [22]. The animal under such condition has less water intake, which results in poor growth [23]. There are not enough literature to debunk this variability or established how true this variable as influenced by season, but this may be due to reflects changes in the average state of the micro-climatic atmosphere in relative with the air speed velocity, and photoperiodic of the rabbit house [24].

The average values for TEMP, RH, and THI per season are shown in **Table 1**. The estimated values for TEMP were 34.78 ± 0.17, 33.41 ± 0.22, 31.17 ± 0.22, and 33.55 ± 0.19°C for late dry, early rain, late rain, and early dry season, respectively ($p < 0.05$), and the corresponding RH values were 56.89 ± 0.17, 57.23 ± 0.16, 59.03 ± 0.25, and 57.17 ± 0.20% ($p < 0.05$). For THI, the reported values were 32.03 ± 0.14 in late dry, 30.88 ± 0.19 in early rain, 29.03 ± 0.19 in late rain, and 30.99 ± 0.16 in early dry ($p < 0.05$). According to Marai et al. [19], late rain season is considered as severe heat stress season (29.0–30.0), and late dry, early, and early dry seasons are considered very severe heat stress seasons (>30.0).

**Table 2** shows the descriptive statistics for doe productivity performance by season THI. The late dry season and the first part of early rain season displayed the worse values for conception rate (CR), from 0 to 20 and 16%, respectively. Therefore, between 100 and 80% of the matings did not lead to a delivery. Kindling interval (KI) was very high at the first part of early rain season (98 days). The reason for this large mating failure rate and high KI could be attributed to heavier

| | Late dry | Early rain | Late rain | Early dry |
|---|---|---|---|---|
| RH (%) | 56.89 ± 0.17$^{d}$ | 57.23 ± 0.16$^{b}$ | 59.03 ± 0.25$^{a}$ | 57.17 ± 0.20$^{bc}$ |
| TEMP (°C) | 34.78 ± 0.17$^{a}$ | 33.41 ± 0.22$^{bc}$ | 31.17 ± 0.22$^{d}$ | 33.55 ± 0.19$^{b}$ |
| THI | 32.03 ± 0.14$^{a}$ | 30.88 ± 0.19$^{bc}$ | 29.03 ± 0.19$^{d}$ | 30.99 ± 0.16$^{b}$ |

*Values are least square means (±SEM).*
$^{abcd}$*Means within the same row having different superscripts are significantly ($p < 0.05$) different.*

**Table 1.**
*Relative humidity (%), temperature (°C), and temperature-humidity of the experimental season.*

| Season | Parity | N | SMD | CR (%) | KI (days) | GK |
|---|---|---|---|---|---|---|
| Late dry | 1st | 50 | 10 | 20 | — | 20 |
| | 2nd | 50 | 4 | 0 | — | 0 |
| Early rain | 3rd | 50 | 4 | 16 | 98 | 1 |
| | 4th | 50 | 17 | 68 | 49 | 49 |
| Late rain | 5th | 50 | 13 | 52 | 49 | 29 |
| | 6th | 50 | 14 | 56 | 49 | 34 |
| Early dry | 7th | 50 | 13 | 52 | 49 | 26 |
| | 8th | 50 | 17 | 68 | 49 | 18 |

*SMD, successive mated doe; CR, conception rate; KI, kindling interval; GK, group of kits that reached sexual maturity at 22 weeks.*

**Table 2.**
*Productivity performance of does within the expected parities as influenced by season THI.*

| Parameter | Index |
|---|---|
| Weaned/doe/year | 27.69 |
| Slaughtered/doe/year | 23.54 |
| Live weight/doe/year | 43.54 |
| Dressed weight/doe/year | 24.38 |

**Table 3.**
*Doe productivity indices during the period of experiment.*

precipitation and higher temperature (38–42°C) during this period. These findings agree with those of Isaac et al. [25], who reported that onset of rain will superimpose deleterious reductive effect on productive potential of breeding does. The remaining seasons showed similar range of values for CR (68% for the last part of early rain, from 52 to 56% for late rain, and from 52 to 68% for early dry), and the KI exhibited a value of 49 days in all them. Akpo et al. [26] reported higher values for KI in the first three parities (51–65 days) and similar values to ours from fourth parity (44 days). The highest value of 49 groups of kits that reached the maturity of 22 weeks was kindled in the last month (June) of the early rain with 29.82 THI (**Figure 3**), proceeding with most favorable 3 months of the late rain season with the severe heat stress on the rabbit body metabolic mechanism. This rabbit's grouped offspring was raised under severe heat stress and therefore could have developed favorable metabolic mechanisms to adapt to heat stress.

**Table 3** shows the productivity indices under tropical conditions. Rabbit females could reach to 27.69 weaned/year, 23.57 slaughtered/year, 43.54 kg of live weight/year, and 24.38 kg of dressed weight/year. These results agree to Lebas et al. [10] who recorded at least 30 weaned/doe/year in the tropics under identical production conditions of semi-intensive reproductive rate system.

## 4. Conclusion

Tropical rabbit farming was subjected to severe heat stress in the late dry, early rain, and early dry seasons. The most favorable season was late rain with the lowest THI and the highest productivity potential parameters.

## Author details

Dennis Osarenren Asemota* and Arierhire Michael Orheruata
Department of Animal Science, University of Benin, Benin City, Nigeria

*Address all correspondence to: aseadonis@yahoo.com

## References

[1] Parker A, Wilfred A, Hidell T. Environmental monitoring: The key to effective sanitation. Laboratory Animals. 2003;**32**:26-29

[2] García-Ispierto I, López-Gatius F, Santolaria P, Yániz JL, Nogareda C, López-Béjar M. Factors affecting the fertility of high producing dairy herds in northeastern Spain. Theriogenology. 2007;**67**:632-638

[3] Igono MO, Jotvedt G, Sanford-Crane HT. Environmental profile and critical temperature effects on milk production of Holstein cows in desert climate. International Journal of Biometeorology. 1992;**36**:77-87

[4] LPHSI. Livestock and Poultry Heat Stress Indices Agriculture Engineering Technology Guide. Clemson, SC, USA: Clemson University; 1990

[5] Marai IFM, Ayytat MS, Abd el-Monem UM. Growth performance and reproductive traits at first parity of New Zealand white female rabbits affected by heat stress and its alleviation under Egyptian conditions. Tropical Animal Health and Production. 2001;**33**:451-462

[6] Fernandez CI, Blas E, Concha C. Growth and some carcass traits of adult rabbits under high ambient temperature. World Rabbit Science. 1994;**2**:147-151

[7] Marai IFM, Abd El-Samee AM, El-Gafaary MN. Criteria of response and adaptation to high temperature for reproductive and growth traits in rabbits. Options Mediterraneennes, Seree A. 1991;**17**:127-134

[8] Marai IFM, El-Kelawy HM. Effect of heat stress on the reproduction in females of rabbits. In: Proceedings of 1st International Conference on Indigenous versus Acclimatized Rabbits; El-Arish, North Siani, Egypt. 1999

[9] McNitt JI, Steven DL, Peter RC, Nephi MP. Rabbit Production. 2000. Available from: www.amazon.com/Rabbit-production-Cached

[10] Lebas F, Courdet P, de Rochambean H, Thebauit RG. The rabbit husbandry, health and production. FAO Animal Production and Health Series. 1997;**21**:102-155

[11] Olson LC, Palotay JL. Epistaxis and bullae in cynomolgus macaques (*Macaca fascicularis*). Laboratory Animal Science. 1983;**33**:377-379

[12] Jessen C. Temperature Regulation in Human and Other Mammals. New York: Springer; 2001

[13] Fayez I, Marai M, Alnaimy A, Habeeb M. Thermoregulation in rabbits. Options Méditerranéennes. 1994;**8**:33-41

[14] Choudhary H, Goswami RN, Das D, Das A, Roycoudhury R. Genetic studies on the reproductive performance of Soviet Chinchilla breed of rabbit under the agroclimatic conditions of North Eastern region. The Indian Journal of Animal Sciences. 2001;**71**:946-949

[15] Oguike MA, Okocha NL. Reproductive performance of rabbits remated at different intervals post-partum. African Journal of Agricultural Research. 2008;**3**(6):412-415

[16] Meteorological Section of the Nigeria Airports Authority-NAA. Edo State report; 2013

[17] Smith E, Stockwell JD, Schweitzer I, Langley SH, Smith AL. Evaluation of cage microenvironment of mice housed on various types of bedding materials. Contemporary Topics in Laboratory Animal Science. 2004;**43**:12-17

[18] Colin M, Lebas F. Rabbits of the World (in French). Lempdes, France:

Association Francaise de cuniculture edit; 1995. p. 330

[19] Fayez I, Marai M, Ayyat MS, Abd El-Monem UM. Young doe rabbit performance traits as affected by dietary zinc, copper, calcium or magnesium supplements under winter and summer conditions of Egypt. In: Proceedings of 7th World Rabbit Congress; Valencia, Spain. 2000. pp. 313-320

[20] Odubote IK, Ohiosimuan OO, Oseni SO. The potentials of rabbit as a component of peri-urban livestock production in Nigeria. In: Zessin KH, editor. Livestock Production & Diseases in the Tropics, Proceedings of the VIII International Conference of the Association of Institutions of Veterinary Medicine; Berlin, Germany. Vol. 1. 1995. pp. 233-237

[21] Fayez I, Marai M, Habeeb AAM, Gad AE. Rabbits productive, reproductive and physiological performance traits as affected by heat stress: A review. Livestock Production Science. 2002;**78**:71-90

[22] Farooq U, Samad HA, Shehzad F, Qayyum A. Physiological responses of cattle to heat stress. World Applied Sciences Journal. 2010;**8**(Special Issue of Biotechnology and Genetic Engineering):38-43

[23] Rowlinson P. Adapting livestock production systems to climate change—Temperate zones. In: Rowlinson P, Steele M, Nefzaoui A, editors. Livestock and Global Change. Proceedings of an International Conference; Hammamet, Tunisia. Cambridge, UK: Cambridge University Press. 2008. pp. 61-63

[24] ACM. 2008. Available from: www.scibd.com/../Ag-Guide-3rd-ed

[25] Isaac LJ, Eko PM, Eko JS, Ekanem E, Essien GB. Effect of breed on performance of rabbits in feed. In: Proceedings of the 35th Annual Conference of the Nigerian Society for Animal Production; 14-17 March; University of Ibadan, Nigeria. 2010. pp. 18-19

[26] Akpo Y, Kpodekon TM, Tanimomo E, Djago AY, Youssao AKI, Coudert P. Evaluation of the reproductive performance of a local population of rabbits in South Benin. In: Proceedings of the 9th World Rabbit Congress; 10-13 June 2008; Verona, Spain. 2008

Chapter 5

# The Genetic Improvement in Meat Rabbits

*María-Luz García and María-José Argente*


## Abstract

Rabbits are raised for many different purposes, such as breeding stock for meat, wool and fur, as an educational and experimental animal model, and as pets and show animals. However, this species is main used for meat production. France, Italy and Spain have an important role in the increase of world rabbit meat production through the development of selection programs in this species. Genetic improvement programs have based on development of maternal lines to improve prolificacy and paternal lines to improve growth rate, but the alternative development of multi-purpose lines for litter size and growth traits will be discussed. In this chapter, the variance components of these traits, the response to selection and the main commercial available lines will be reviewed. Universities and public research centers have played a leading role in the development of these lines and in the diffusion of this genetic material through a pyramid scheme from selection nuclei to farmers. Recently, others functional traits are emerging successfully as selection criteria in breeding programs such as ovulation rate, prenatal survival, longevity, feed efficiency, meat quality, uniformity in production, and resistance to digestive disorders.


**Keywords:** average daily gain, feed conversion rate, heritability, litter size, selection

## 1. Introduction

Rabbits are raised for many different purposes, such as breeding stock for meat, wool and fur, as an educational and experimental animal model, and as pets and show animals. However, this species is used mainly for meat production. China and Mediterranean countries concentrate 78% of world production meat [1]. It must note highlighting the leadership of France, Italy and Spain in development of the rabbit selection programs, which have been key to enhance the efficiency in meat production.

## 2. Economic important traits in meat rabbits

The selection objectives in breeding programs are established according to the economic importance of the traits. Economic weights in rabbit meat production have been estimated in different markets, such as the Spanish [2, 3], Australian [4] and French one [5], and in all these studies, the litter size and the feed conversion

| Traits | Unit | Spain | | Australia | France |
|---|---|---|---|---|---|
| | | [2] | [3] | [4] | [5][a] |
| | *Reproductive traits* | | | | |
| Litter size | Increase by 1 | 16.90 | 15.66 | 15.03 | 45.52 |
| Lactation survival | Increase by 1% | 1.96 | 1.71 | 1.70 | |
| Replacement rate per does and year | Increase by 1% | −0.45 | −0.29 | −0.23 | |
| | *Growth traits* | | | | |
| Daily feed intake during fattening | Decrease by 1 g/d | 0.41 | 0.50 | 0.49 | |
| Daily gain during fattening | Increase by 1 g/d | 1.53 | 1.33 | 1.23 | 11.82 |
| Feed conversion rate during fattening | Decrease by 0.1 g/g | 18.80 | 20.19 | | 10.26 |
| *Healthy* | | | | | |
| Resistance to enterocolitis | | | | | 4.41 |

[a]*Economic weights estimated in a context of restricted feeding.*

**Table 1.**
*Economic weights of the main traits of the profit function in €/unit of the trait.*

rate have been reported as the most important traits for rabbit industry (see **Table 1**). The growth rate is easier and cheaper to record than feed conversion rate and has a favourable genetic correlation with it [6]. For this reason, rabbit commercial schemes are based on three-way cross. Two selected lines for litter size at birth or at weaning are crossed to create a commercial doe [7–13], which is mated with a terminal sire from other selected line for growth rate post-weaning or for body weight at a point close to market age [14–17]. The aim of the cross between the maternal lines is to exploit advantage of the expected positive heterosis in reproductive traits, the possible complementarity among the lines and the dispersion of the inbreeding accumulated within the lines [8].

## 3. Genetic parameters for litter size and growth traits

Genetic progress in the selection programs depends mainly on the heritability of the selected trait and on the selection intensity. In this section, a review of quantitative genetic components for litter size and growth traits will be carried out. For litter size at birth, the estimates of the heritability show in general low values (0.05 to 0.20 and 0.11 on average) and tended to decrease slight from birth to slaughter (0.00 to 0.13 and 0.08 on average for number born alive, 0.02 to 0.12 and 0.07 on average for litter size at weaning, and 0.06 to 0.08 and 0.07 on average for litter size at slaughter, see **Table 2**). The estimates of the ratios of permanent environmental variance to the phenotype variance are also rather low for litter size at birth. In agreement to heritability, the estimated values decrease from birth (0.11 on average) to market time (0.08 on average). These findings are an indication of high effect of environmental influence on litter size and the low repeatability. Regarding genetic correlations between litter size traits, the estimates present positive and high values, ranging from +0.96 to +0.99 for litter size at birth and number born alive, and from +0.60 to +0.98 for number born alive and litter size at weaning [11, 24, 33].

For growth traits, there are many estimates of heritability for weaning and slaughter weight (see **Table 3**). The average values of these estimates are moderate (0.18 for weaning weight and 0.22 for slaughter weight). However, these estimates

| LS | | NBA | | NW | | NS | | Line/breed | References |
|---|---|---|---|---|---|---|---|---|---|
| $h^2$ | $p^2$ | $h^2$ | $p^2$ | $h^2$ | $p^2$ | $h^2$ | $p^2$ | | |
| 0.20 | 0.25 | | | 0.09 | 0.12 | | | New Zealand White | [18] |
| 0.10 | 0.07 | 0.07 | 0.09 | 0.07 | 0.07 | 0.07 | 0.06 | Line selected by OR and LS | [19] |
| 0.10 | 0.09 | | | | | | | Environmental Variance of LS | [20] |
| 0.11 | 0.08 | 0.10 | 0.09 | 0.09 | 0.07 | | | Line A | [10, 21, 22] |
| 0.08 | 0.12 | 0.05 | 0.09 | 0.02 | 0.07 | | | Line H | [21, 22] |
| | | 0.09 | 0.10 | 0.08 | 0.08 | | | Line LP | [21] |
| | | 0.09 | 0.11 | 0.07 | 0.13 | | | Line R | [21, 22] |
| 0.18 | 0.09 | 0.07 | 0.10 | 0.05 | 0.08 | 0.05 | 0.07 | Line V | [23] |
| 0.13 | 0.05 | 0.05 | 0.09 | 0.02 | 0.06 | | | ITELV2006 line | [24] |
| | | 0.05 | 0.09 | | | | | Pannon White | [25] |
| 0.13 | 0.10 | 0.00 | 0.06 | | | | | Pooled Poured Breed | [26] |
| 0.05 | 0.09 | 0.03 | 0.09 | | | | | Brazilian Synthetic Line | [27] |
| 0.12 | 0.06 | 0.09 | 0.07 | | | | | Pannon Ka | [28] |
| 0.05 | 0.11 | 0.07 | 0.11 | | | | | Pannon Large | [29] |
| 0.07 | 0.10 | 0.07 | 0.09 | | | | | Pannon White | [30] |
| 0.11 | 0.09 | 0.08 | 0.08 | 0.06 | 0.03 | | | Line Prat | [10] |
| 0.09 | 0.21 | 0.12 | 0.20 | 0.09 | 0.16 | 0.07 | 0.12 | Local line | [31] |
| 0.19 | 0.19 | | | 0.08 | 0.19 | | | Danish While | [32] |

*OR: ovulation rate.*

**Table 2.**
*Heritability ($h^2$) and permanent effect ($p^2$) of litter size at birth (LS), number of kits born alive (NBA), number of kits at weaning (NW) and number of rabbits at slaughter (NS).*

present widely range of values (0.03 to 0.48 for weaning weight and 0.06 to 0.67 for slaughter weight); that can be related to different weaning age, from 28 days in semiintensive management to 42 days of age in extensive management, and different slaughter time, from 9 week in Spain to 13 weeks of age in Italy (see review by [46]). Contrarily, the estimates of heritability for growth rate show a narrow range (0.12 to 0.34) and moderate average value (0.22). A reduced number of studies has been carried out to analyse the genetic determination of feed conversion rate (see **Table 3**). The average value of heritability for feed conversion rate is similar those of growth rate (0.29), varying in a small range such as growth rate (0.22 to 0.42). The litter effect is especially important for weaning weight (0.47 on average), and in lesser extent for slaughter weight (0.28 on average), growth rate (0.19 on average) and feed conversion rate (0.12 on average). Some studies have also estimated maternal genetic effects for growth traits. Maternal heritability seems to be slightly higher for weaning weight (0.17 on average) than for slaughter weight (0.10 on average). There is only one estimation for growth rate (0.21), and no estimate has found for feed conversion rate in bibliography. In general, maternal genetic effects are much lesser important than litter effects.

Regarding genetic correlations between growth traits, weight at weaning is positive and highly correlated with weight at slaughter in agreement with [24, 33], ranging from +0.61 to +0.74. Genetic correlation between growth rate and weight at slaughter is higher than at weaning (+0.56 vs. +0.31 [33, 47]). Genetic

| WW | | | | SW | | | | ADG | | | | FCR | | Line/Breed | Reference |
|---|---|---|---|---|---|---|---|---|---|---|---|---|---|---|---|
| $h^2$ | $c^2$ | $p^2$ | $h^2_m$ | $h^2$ | $c^2$ | $p^2$ | $h^2_m$ | $h^2$ | $c^2$ | $p^2$ | $h^2_m$ | $h^2$ | $c^2$ | | |
| | | | | 0.41 | | | | | | | | | | New Zealand White | [34] |
| 0.06 | 0.43 | | | 0.13 | 0.26 | | | 0.19 | 0.13 | | | 0.22 | 0.10 | ConsoResidual line | [35] |
| 0.04 | 0.33 | | | 0.11 | 0.22 | | | 0.22 | 0.17 | | | 0.23 | 0.16 | ADGrestrict line | [35] |
| 0.15 | | 0.27 | | 0.19 | | 0.14 | | 0.21 | | 0.10 | | | | Line B | [36] |
| 0.15 | | 0.18 | | 0.15 | | 0.12 | | 0.17 | | 0.10 | | | | Line R | [36] |
| 0.03 | 0.64 | 0.07 | | 0.06 | 0.38 | 0.08 | | | | | | | | ITELV2006 line | [24] |
| | | | | 0.20 | | a | | 0.25 | | a | | | | Pannon White | [37] |
| | | | | | | | | 0.27 | 0.14 | | | | | Pannon White | [38] |
| 0.04 | 0.72 | | | 0.12 | 0.51 | | | 0.17 | 0.40 | | | | | Line selected by body weight at 70 d | [15] |
| 0.48 | | | 0.25 | 0.39 | | | 0.11 | | | | | | | Brazilian Synthetic Line | [27] |
| 0.08 | 0.44 | | 0.18 | 0.08 | 0.26 | | 0.05 | | | | | | | Brazilian Synthetic Line | [27] |
| 0.24 | 0.31 | 0.01 | | 0.17 | 0.18 | 0.18 | 0.00 | | | | | | | Angora line | [39] |
| 0.09 | 0.35 | 0.11 | | 0.13 | 0.28 | 0.05 | | 0.14 | 0.27 | 0.01 | | | | Line selected by OR and LS | [40] |
| 0.41 | | a | | 0.37 | | a | | 0.34 | | a | | | | Line Prat | [33] |
| | | | | | | | | 0.21 | 0.17 | | | 0.25 | 0.22 | Line Prat | [6] |
| | | | | | | | | 0.21 | 0.12 | | | 0.32 | 0.07 | Line Caldes | [41] |
| | | | | | | | | 0.17 | 0.32 | | | | | Danish White | [42] |
| 0.42 | | 0.18 | 0.09 | 0.27 | | | 0.27 | 0.21 | | | 0.21 | | | New Zealand White | [43] |
| 0.25 | 0.44 | | | 0.24 | 0.22 | | | 0.22 | 0.09 | | | 0.33 | 0.07 | AGP39 | [44] |
| 0.12 | 0.52 | | | 0.14 | 0.27 | | | 0.12 | 0.12 | | | 0.42 | 0.07 | AGP59 | [44] |
| 0.09 | 0.52 | | | 0.67 | 0.26 | | | 0.41 | 0.21 | | | 0.27 | 0.17 | Divergent lines for residual feed efficiency | [45] |

*OR: ovulation rate. LS: litter size. a: effect included in the model but not display.*

**Table 3.**
*Heritability ($h^2$), common litter effect ($c^2$), permanent effect ($p^2$) and genetic maternal effect ($h^2_m$) of weaning weight (WW), slaughter weight (SW), average daily gain (ADG) and feed conversion rate (FCR).*

correlation between growth rate and feed conversion rate is negative and moderate (−0.4 to −0.5 [6, 35]). The bibliography is scarce and contradictory for genetic correlations between litter size traits and growth traits. There are high and negative estimates between litter size and weight at weaning (−0.85, −0.92 and −0.85 for litter size at birth, number born alive and litter size at weaning, respectively [24]) and estimates close to zero (−0.05, −0.07 and −0.25 for litter size at birth, number born alive and litter size at weaning, respectively [33]). Indeed, it was reported that increases in litter size resulted in a decrease of individual weight at weaning [48, 49]. The genetic correlations between litter size traits with weight at slaughter (+0.11, +0.03 and −0.16 for litter size at birth, number born alive and litter size at weaning, respectively [33]) and growth rate (+0.04, −0.06 and −0.16 for litter size at birth, number born alive and litter size at weaning, respectively [33]) show also values close to zero.

Selection is more complicated for litter size traits than for growth traits. This complexity is due to the fact that the litter size traits display a low heritability and only express in the does, and consequently selection intensity is lower than when both sexes express the trait [12, 50]. In order to increase the accuracy in estimates of genetic values, and therefore the progress into selection program, it is recommended considering as many individual and relative records as possible for genetic evaluation of the does and males, even though generational interval increases [51]. Selection for average daily gain from weaning to slaughtering has been used traditionally as selection criterion to improve of feed conversion rate thus far, since this trait has a moderate heritability and it is lesser affected to common litter effects than the individual weight at specific age (**Table 3**). Moreover, it is much easier and cheaper to measure than feed conversion rate and it has a negative favourable correlation with it [6, 35]. However, the development cheap electronic devices nowadays that enable recording of individual feed intake in this species, together moderate heritability of this trait and its moderate genetic correlation with average daily gain (−0.4 to −0.5), have challenged whether selection for average daily gain is the best way to improvement of feed efficiency, instead of direct selection (see review [46]).

## 4. Selected lines

Traditionally, rabbit commercial schemes have based on development of specialised lines to improve prolificacy (maternal lines) and to improve growth rate (paternal line) as it was commented in Section 2 [7–17]. However, the foundation and development of specialised lines is an activity with the high requirements, organisation, experience, and money needed, that not all countries can carry out. In countries where the rabbit industry has not yet reached a proper level of organisation, it may not be appropriate to select dam and sire lines for a subsequent cross-breeding program [52]. An alternative could be the development of multi-purpose lines, through simultaneous selection for litter size and growth traits [27].

In maternal lines, the most common direct criteria used in selection programs is litter size at birth or at weaning (see **Table 4**). Although, litter size at weaning show a lower heritability than litter size at birth (see **Table 2**); the majority of maternal lines are selected by litter size at weaning, since this trait reflects both the prolificacy as well as the maternal ability of the doe (**Table 4**). In some commercial lines, the selection criterium is weight at weaning, a trait relates to the ability of the doe for lactating and nourishing the progeny [56]. The response due to selection in these maternal lines has ranged between 0.05-0.13 kits born alive or weaned per litter and generation [8].

| Name | Country | Origen | Selection criteria | Number of generations | Reference |
|---|---|---|---|---|---|
| INRA2066 | France | Californian & Giant Himalayan | Litter size at birth | More than 34 generations | [53] |
| INRA2666 | France | INRA2066 & Line V | Litter size at weaning | Since 1999 | [54] |
| INRA9077 | France | New Zealand White & Bouscat White | Litter size at birth | Since 1998 | [55] |
| INRA1777 | France | INRA1077 | Litter size at birth & individual weaning weight | More than 5 generations | [56] |
| Line A | Spain | New Zealand White | Litter size at weaning | More than 44 generations | [57] |
| Line V | Spain | Four specialised maternal lines | Litter size at weaning | More than 39 generations | [57] |
| Line H | Spain | Hyperprolific commercial does | Litter size at weaning | More than 22 generations | [57] |
| Line LP | Spain | Long-lived commercial does | Litter size at weaning | More than 8 generations | [57] |
| Line PRAT | Spain | A closed population with crossbred animals | Litter size at weaning | Since 1992 | [58] |
| Pannon Ka | Hungary | Crossbreds & Pannon White | Number of kits born alive | Since 1999 | [59] |
| APRI | Egypt | Baladi Red & Line V | Litter weight at weaning | Since 2002 | [60] |
| ITELV2006 | Argelia | INRA2666 and local population | Litter size at birth and body weight at 75 days | Since 2003 | [61] |
| Uruguay NZW | Uruguay | New Zealand White | Litter size at weaning | More than 5 generations | [62] |
| Uruguay V | Uruguay | Line V | Litter size at weaning | More than 5 generations | [62] |

**Table 4.**
*Maternal lines for meat rabbit production.*

In paternal lines, in order to improve feed conversion rate as comment before, the most common direct criteria used in selection programs is postweaning daily gain from weaning to slaughtering. Other selection criteria used in paternal lines are those related to the weight at slaughter (see **Table 5**). Recently, residual feed intake was investigated experimentally as a direct way to improve the feed conversion rate [44, 45, 48]. The response to selection in paternal lines range between 18 and 35 g/generation for weight at slaughter and between 0.45 and 1.23 g/d generation for daily gain, with positive correlated response on adult weight and feed intake and negative correlated response on feed conversion, dressing percentage and maturity at a fixed weight [8, 48].

In multi-purpose lines, both growth and reproductive traits are selected (**Table 6**). Thus, there are lines selected simultaneously by individual weight at slaughter and litter size traits, and by thigh muscle volume (TMV) measured on computer tomography (CT) and litter weight or average daily gain. The problem of selection by TMV is the high costs and the long generation intervals [59].

The oldest program for rabbit breeding and improvement is the French program that was started in 1969 by French National Institute for Agricultural Research (INRA-SAGA, Toulouse), and followed by the Spanish programs that started in

| Name | Country | Origen | Selection criteria | Number of generations | Reference |
|---|---|---|---|---|---|
| Line R | Spain | Two paternal lines | Postweaning daily gain | More than 32 generations | [57] |
| Line Caldes | Spain | Crossbreds | Postweaning daily gain | Since 1992 | [63] |
| Italian Silver | Italy | Argenté de Champagne | Postweaning daily gain | Since 2000 | [57] |
| ALEX | Egypt | Baladi Black & Line V | Postweaning daily gain | More than 7 generations | [13] |
| Altex | USA | ¼ California & ¼ Giant Himalayan & ½ Flemish Giat | Individual weight at 70 days | Since 1994 | [15] |

**Table 5.**
*Paternal lines for meat rabbit production.*

| Name | Country | Origen | Selection criteria | Number of generations | Reference |
|---|---|---|---|---|---|
| INRA1077 | France | New Zealand White & Bouscat White | Litter size at birth & Individual weight at 63 days | More than 30 generations | [64] |
| Giante de España | España | Flemish Giant & Lebrel Español | Litter weight at weaning & growth rate during fattening | Since 1984 | [65] |
| Italian New Zealand White | Italy | New Zealand White | Litter size at 21 days & Individual weight at 60 days | Since 1980 | [66] |
| Italian California | Italy | California | Litter size at 21 days & Individual weight at 60 days | Since 1980 | [66] |
| Pannon White | Hungary | New Zealand White & California | Litter weight at 21 days & Thigh muscle volume | Since 2010 | [59] |
| Pannon Terminal L | Hungary | Crossbreds & Pannon White | Postweaning daily gain & Thigh muscle volume | Since 2005 | [59] |
| Moshtohor | Egypt | Sinai Gabali & Line V | Litter weight at weaning & individual weight at 56 days | Since 2006 | [13] |
| Saudi-3 | Saudi Arabia | Saudi Gabali & Line V | Litter weight at weaning and weight at 84 days | Since 2000 | [13] |
| Botucatu | Brazil | Norfolk line | Litter size at weaning & Postweaning daily gain | Since 1998 | [27] |

**Table 6.**
*Multi-purpose line for meat rabbit production.*

1976 for the Department of Animal Science at Universitat Politècnica de València (UPV, Valencia) and in 1992 for Rabbit Science Unit at Institute of Agrifood Research and Technology (IRTA). The INRA-SAGA has developed several maternal lines as INRA2066, INRA2666, INRA1777 and INRA9077, and a synthetic multi-purpose line as INRA1077. In Spain, the UPV and IRTA have created the maternal lines A, V, H, LP and PRAT and the paternal lines R and Caldes. Besides, University of Zaragoza has developed a multi-purpose line Gigante de España [57].

Other selection programs in rabbits have been carried out both inside and outside Europe. For example inside Europe, Kaposvár University in Hungary has developed the maternal line Pannon Ka and multi-purpose lines Pannon White and Pannon Terminal L, and two cooperative centres from Emilia-Romagna in Italy have created the paternal line Italian Silver and the multi-purpose lines Italian New Zealand White and California. Outside Europe, we can found the maternal lines APRI (at the Animal Production Research Institute in Egypt), ITEL2066 (at the Institut Technique de l'Elevage -ITELV- and at Tizi Ouzou University in Algeria), and Uruguay NZW and V (at Instituto Nacional de Investigaciones Agropecuarias of las Brujas in Uruguay), and the paternal lines ALEX (at Alexandria University in Egypt) and Altex (at Texas A&M University in USA) as well as the multi-purpose lines Moshtohor (at Benha University in Egypt), Saudi-3 (at King Saud University in Saudi Arabia) and Botucatu (at Faculdade de Medicina Veterinária e Zootecnia of Botucatu in Brazil). It must note that most of the lines developed outside Europe have had the collaboration of the UPV and INRA-SAGA. Furthermore, the rabbit farmer can also purchase in market animals from the maternal and paternal lines from several private companies, mainly French and Spanish as Eurolap Hyla, Grimaud Frères Sélection, Hycole, Hypharm, and Granja Jordán among others.

## 5. Selection experiments

New traits are emerging as criterium selection in breeding programs, both maternal lines and parental lines. Accordingly, selection experiments have been carried out in different rabbit populations. Different strategies have been adopted for estimating the genetic progress in these experiments, as the using divergently selected lines or the using a control population. Divergent selection allows us to use each line as control of the other, but estimated response can be biased when response is no symmetry in both lines. Control population provides an unbiased estimate of response to selection since working with non-selected animals from the same population. Selection for ovulation rate, prenatal survival, longevity, feed efficiency, meat quality, uniformity in production, and resistance to Pl digestive disorders has been reviewed in this section.

### 5.1 Selection for ovulation rate and prenatal survival

Selection for ovulation rate and prenatal survival has been proposed as an indirect approach for increasing litter size since these parameters limit it. In turn, uterine capacity limits prenatal survival, thus its selection has been postulated in order to improve litter size [67]. There has been carried out one selection experiment for ovulation rate [68], two divergent selection experiment for uterine capacity (one in UPV [69] and other in INRA-SAGA [70]), and one two-step selection experiment for ovulation rate and litter size [71]. The estimated response to selection for ovulation rate using a control population was 0.21 ova per generation without any correlated response in litter size, as consequence a reduction in fetal survival [68]. The difference between the divergent lines for uterine capacity showed that selection was effective for uterine capacity and a correlated response was found in embryo survival in the experiment of UPV [72] and in fetal survival in the experiment of INRA-SAGA [70]. An asymmetric correlated response in litter size was reported after 10 generation of selection in UPV experiment using a control population; whereas increasing uterine capacity was not accompanied by a correlated response in litter size, decreasing it reduced litter size by 0.19 kits per generation because of lower embryo and fetal survival [73]. Two-stage selection by ovulation rate and litter size has successful and showed a correlated response in litter size by 0.12 kits per generation [71].

### 5.2 Selection for longevity

Due partially to negative correlated response to high selection for production on voluntary culling in dam, the longevity has been proposed as new selection objective in breeding programs in rabbits. In this sense, two selection experiments have been performed to improve longevity: one in the UPV and other in the INRA-SAGA. The UPV's experiment has allowed to create the LP line. This line was founded by selecting females from commercial farms with extremely high number of parturitions (between 25 and 41 parities) and a constraint on prolificacy (from 7.5 to 11.9 young born alive) [74]. Once the LP line was constituted, the selection is being carried out by litter size at weaning and this line is currently in 17th generation. The INRA-SAGA has performed a divergent selection experiment for longevity. The selection criterium was the total number of artificial inseminations after the first parity [75]. Both experiments have showed a favourable correlated response on doe's body reserves. However, response to longevity has been limited, due to this trait has a small heritability and the time required obtaining pertinent information is long.

### 5.3 Selection for feed efficiency

Feed efficiency has been traditionally measured as feed conversion rate, i.e., the ratio between feed intake and body weight gain over a fixed range of days. More recently, residual feed intake has emerged as new trait for improving of feed efficiency. However, residual feed intake is no ease to measure, since to require using equations in order to estimate the difference between actual feed intake and expected feed intake according to the requirements for the maintenance and the growth of the animal. Several divergent selection experiments in rabbits for feed conversion rate [76] and residual feed intake [35, 45, 77] have been carried out. The divergent selection experiment of Moura et al. [76] reports a difference between lines, having the high line lower feed conversion rate than the low one at the end of the experiment. The estimated response to selection using mixed model technique was 0.6% per generation. The divergent selection experiment on residual feed intake of Larzul and de Rochambeau [45] only had one generation of selection, nothing can be said about whether selection was successful since the difference between the lines was not significant. The experiment of selection for residual feed intake between 30 and 65 d of age of Drouilhet et al. [35, 77] showed a decreasing in residual feed intake of 0.9% per generation (−39 g), and a correlated response of 0.8% (−0.20 g) in feed conversion rate after nine generations. No correlated response was found for growth rate, showing that selection acted upon reducing appetite [78, 79].

### 5.4 Selection for quality meat

Intramuscular fat is a main meat quality factor, since affecting sensory properties and the nutritional value of the meat. A divergent selection experiment on intramuscular fat in muscle *Longissimus dorsi* was carried out by Zomeño et al. [80]. After seven generations of selection, the divergence between lines was around 5% per generation (1.09 g/100 g), with a symmetrical response [81]. There were no correlated responses in pH and in colour and in any sensory attributes [82]. A positive correlated response was found on fat in *Biceps femoris*, in *Supraspinatus* and *Semimembranosus proprius* muscles, and in perirenal fat content, which was greater in the high line [83]. An increase in dissectible fat leads to deterioration in carcass. However, the amount of dissectible fat in rabbit carcasses is low still (2.5% at 9 weeks and 3.5% at 13 weeks, [84]), in order to consider that selection for intramuscular fat can deteriorate carcass in this species.

### 5.5 Selection for uniformity in production

Uniformity in production is an interesting trait for rabbit industry. Two divergent selection experiments for environmental variability have been carried out one in INRA-SAGA for homogeneity in weight at birth and other in University Miguel Hernández de Elche (UMH) for homogeneity in litter size at birth. The INRA-SAGA's experiment showed a lower within-litter birth weight standard deviation in the Homogeneous line than in the Heterogenous line after 10 generations (7.34 g vs. 11.26 g [85]). Moreover, the Homogeneous line exhibited higher litter size at weaning and lower mortality at birth and at weaning than the Heterogeneous line. No correlated response was reported for the individual weight at birth or the standard deviation and individual weight at weaning [86]. A higher homogeneity in weight birth within litter was related to higher length and capacity of the uterine horn, thus the divergence between the lines could be at least partly due to their characteristics of the reproductive tract [87]. In the experiment of UMH, after 10 generations of selection, the environmental litter size variance was 2.7 $kits^2$ in the Homogeneous line and 4.4 $kits^2$ in the Heterogeneous line [88]. A low variability in litter size in the Homogeneous line was related to better adaptation to environment with less response to stress and diseases, i.e. with does more resilient [89]. Therefore, decreasing litter size variability can favour the dam's survival in the farm. Moreover, selection for litter size variability shows a negative response correlated to litter size, i.e., a reduction in litter size variability was accompanied by a larger litter size at birth [88]. A higher litter size in the Homogeneous line was related to a higher number of implanted embryos [90], as consequence a higher embryonic development at early gestation in this line [91, 92].

### 5.6 Selection for resistance to digestive disorders

A divergent selection experiment to resistance to enteropathies disorders was performed in INRA-SAGA. A binary score based on the observed signs of enteropathy during the growing period was the selection criterion. The resistance animals showed similar mortality and growth rate to those of sensitivity animals, but cumulative mortality was lower in resistant than sensitivity animals, when animals were inoculated with an enteropathogenic *E. coli* 0103 strain [93].

## 6. Conclusions

Traditionally, rabbit commercial schemes have based on development of specialised lines to improve prolificacy (maternal lines) and to improve growth rate (paternal line). However, not all countries have a proper level organisation, being an alternative the development of multi-purpose lines for litter size and growth. Universities and public research centers have played a leading role in the development of these lines. Litter size and growth rate have traditionally been the selection criteria in the selection schemes for these lines. Recently, others functional traits are emerging strongly as selection criteria in breeding programs such as ovulation rate, prenatal survival, longevity, feed efficiency, meat quality, uniformity in production, and resistance to digestive disorders.

## Acknowledgements

The Project AGL2017-86083-C2-2-P funding by FEDER/Ministerio de Ciencia e Innovación-Agencia Estatal de Investigación and the Project AICO/2019/169 funding by Valencia Regional Government have allow to conduct this chapter.

## Conflict of interest

The authors declare no conflict of interest.

## Author details

María-Luz García and María-José Argente*
Departamento de Tecnología Agroalimentaria, Escuela Politécnica Superior de Orihuela, Universidad Miguel Hernández de Elche, Orihuela, Spain

*Address all correspondence to: mj.argente@umh.es

## References

[1] FAOSTAT. Food and Agriculture Organization of the United Nations [Internet]. 2020. Available from: http://www.fao.org/faostat [Accessed: 2020-05-07]

[2] Armero Q, Blasco A. Economic weights for rabbit selection indices. Journal Applied Rabbit Research. 1992; 15:637-642.

[3] Cartuche L, Pascual M, Gómez EA, Blasco A. Economic weights in rabbit meat production. World Rabbit Science. 2014; 22:165–177. DOI: 10.4995/wrs.2014.1747

[4] Prayaga KC, Eady S. Rabbit farming for meat production in Australia: preliminary estimates of economic values for production traits. Asian-Australian Journal Animal Science. 2000;13: 357–359

[5] Eady SJ, Garreau H. An enterprise gross margin model to explore the influence of selection criteria for breeding programs and changes to management systems. In Proceedings of 9th World Rabbit Congress. 10-13 June; Verona, Italy; 2008. p. 61-66

[6] Piles M, Gómez EA, Rafel O, Ramon J, Blasco A. Elliptical selection experiment for the estimation of genetic parameters of the growth rate and feed conversion ratio in rabbits. Journal of Animal Science. 2004;82:654-660. DOI: 10.2527/2004.823654x

[7] Baselga M, Blasco A. Mejora genética del conejo de producción de carne. Ed. Mundi Prensa, 1989. Madrid, Spain

[8] Baselga M. Genetic improvement of meat rabbits. Programmes and diffusion. In Proceeding of 8th World Rabbit Congress. 7-10 Septembre; Puebla, Mexico. 2004. p. 1-13

[9] Lebas F, Coudert P, Rochambeau H, Thébault RG. The rabbit: husbandry, health and production. FAO Animal Production and Health Series No. 21. Rome, FAO [Internet]. 1997. Available from: http://www.fao.org/docrep/t1690e/t1690e00.HTM [Accessed: 2020-05-07]

[10] Piles M, García ML, Rafel O, Ramon J, Baselga M. Genetics of litter size in three maternal lines of rabbits: repeatability versus multiple-trait models. Journal of Animal Science 2006; 84(9):2309–2315. DOI:10.2527/jas.2005-622

[11] Ragab M., Baselga M. A comparison of reproductive traits of four maternal lines of rabbits selected for litter size at weaning and founded on different criteria. Livestock Science 2011;136:201–206. DOI: 10.1016/j.livsci.2010.09.009

[12] Khalil MH, Al-Saef AM. Methods, criteria, techniques and genetic responses for rabbit selection: A review. In Proceeding of 9th World Rabbit Congress. 10-13 June; Verona, Italy; 2008. p. 1–22

[13] Khalil MH, Bolet G. Sustainable Rabbit Breeding and Genetic Improvement Programs Achieved in Developing Countries. In Proceedings of the World Congress on Genetics Applied to Livestock Production. Leipzig, Germany. 2010. p. 0962

[14] Rochambeau H de, de la Fuente LF, Rouvier R, Ouhayoun J. Sélection sur la vitesse de croissance post-sevrage chez le lapin. Genetic Selection Evolution 1989; 21:527–546. DOI: 10.1186/1297-9686-21-4-527

[15] Lukefahr SD, Odi HB, Atakora JKA. Mass selection for 70-day body weight in rabbits. Journal of Animal Science. 1996;74:1481–1489. DOI:10.2527/1996.7471481X

[16] Piles M, Blasco A. Response to selection for growth rate in rabbits.

World Rabbit Science. 2003;11(2):53–62. DOI: 10.4995/wrs.2003.497

[17] Larzul C, Gondret F, Combes S, Rochambeau H de. Divergent selection on 63- day body weight in the rabbit: response on growth, carcass and muscle traits. Genetic Selection Evolution. 2005;37:105–122. DOI: 10.1186/1297-9686-37-1-105

[18] Ayyat MS, Marai IFM, El-Sayiad GhA. Genetic and non-genetic factors affecting milk production and preweaning litter traits of New Zealand white does under Egyptian conditions. World Rabbit Science. 1995;3(3): 119-124. DOI:10.4995/wrs.1995.250.

[19] Badawy AY, Peiró R, Blasco A, Santacreu MA. Correlated responses on litter size traits and survival traits after two-stage selection for ovulation rate and litter size in rabbits. Animal. 2019;3 (3):453–459. DOI:10.1017/S1751731118002033

[20] Blasco, A., Martínez-Álvaro, M., García, M. Ibañez-Escriche N., Argent MJ. Selection for environmental variance of litter size in rabbits. Genetic Selection Evolution. 2017;49:48. DOI: 10.1186/s12711-017-0323-4

[21] El Nagar AGF. Genetic analysis of longevity in specialized lines of rabbits. [tesis]. Universidad Politècnica de València; 2015. DOI: 10.4995/Thesis/10251/52390

[22] Fernández EN, Sánchez JP, Martínez R, Legarra A, Baselga M. Role of inbreeding depression, non-inbred dominance deviations and random year-season effect in genetic trends for prolificacy in closed rabbit lines. Journal Animal Breeding and Genetic. 2017;134(6):441-452. DOI: 10.1111/jbg.12284

[23] García ML, Baselga M. Genetic response to selection for reproductive performance in maternal line of rabbits. World Rabbit Science. 2002;10(2):71-76. DOI: 10.4995/wrs.2002.478

[24] Ezzeroug R, Belabbas R, Argente MJ, Berbar A, Diss S, Boudjella Z, Talaziza D, Boudahdir N, Garcia ML. 2020. Genetic correlations for reproductive and growth traits in rabbits. Canadian Journal of Animal Science. 2020. DOI: 10.1139/CJAS-2019-0049

[25] Gyovai P, Nagy I, Radnai I, Németh EB, Szendrő ZS. Heritability and genetic trends of number of kits born alive in a synthetic maternal rabbit line. Italian Journal Animal Science 2009;8(3):110-112. DOI: 10.4081/ijas.2009.s3.110

[26] Lukefahr SD, Hamilton HH. Heritability and repeatability estimates of maternal performance traits in purebred and crossbred does. World Rabbit Science. 1997; 5(3): 99-105. DOI: 10.4995/wrs.1997.326

[27] Moura ASMT, Costa ARC, Polastre R. Variance component and response to selection for reproductive, litter and growth traits through a multi-purpose index. World Rabbit Science. 2001;9:77-86. DOI: 10.4995/wrs.2001.449

[28] Nagy I, Farkas J, Curik I, Gorjanc G, Gyovai P, Szendrő Z. Estimation of additive and dominance variance for litter size components in rabbits. Czech Journal of Animal Science. 2014;59: 182-189. DOI: 10.17221/7342-CJAS.

[29] Nguyen TN, Farkas J, Szendrõ Z, Nagy I. Genetic evaluation of litter size traits in Pannon Large rabbits. Animal Science Papers and Reports. 2017; 35(2): 181-192.

[30] Nguyen TN, Farkas J, Szendrõ Z, Nagy I. Genetic Evaluation of Litter Size Traits in Pannon White Rabbits. Agriculturae Conspectus Scientificus. 2017;82 (1):63-67.

[31] Rastogi RK, Lukefahr SD, Lauckner FD. Maternal heritability and repeatability for litter traits in rabbits in a humid tropical environment. Livestock Production Science. 2000;67: 123-128. DOI: 10.1016/S0301-6226(00) 00180-9r

[32] Sorensen P, Kjaer JB, Brenoe UT, Su G. Estimates of genetic parameters in Danish White rabbits using an animal model: II. Litter traits. World Rabbit Science. 2001;9(1):33-38. DOI: 10.4995/ wrs.2001.444

[33] Gomez EA, Rafel O, Ramon J. Genetic relationship between growth and litter size in traits at first parity in a specialised dam line in rabbits. In Proceeding of 6th World Congress on Genetics Applied to Livestock Production. 11-16 January, Armidale, Australia. 1998. p. 552-555

[34] Anous MR. Correlated response of meatiness indicating traits to selection for weight at prevailing marketing age in New Zealand White rabbits. Archiv fur Geflugelkunde. 1999;63:225-228

[35] Drouilhet, H Gilbert, E Balmisse, J Ruesche, A Tircazes, C Larzul, H Garreau. Genetic parameters for two selection criteria for feed efficiency in rabbits. Journal Animal Science. 2013;91(7):3121–3128. DOI: 10.2527/jas.2012-6176

[36] Estany J, Camacho J, Baselga M, Blasco A. Selection response of growth rate in rabbit for meat production. Genetic Selection Evolution. 1992;24:527-237. DOI: 10.1186/1297-9686-24-6-527

[37] Garreau H, Szendrõ Zs, Larzul C, Rochambeau H. de. Genetic parameters and genetic trends of growth and litter size traits in the White Pannon breed. In Proceeding of 7th World Rabbit Congress, 4-7 July, Valence, Spain. 2000. p. 403-408.

[38] Gyovai P, Nagy I, Gerencsér Z, Metzger S, Radnai I, Szendrõ Z. Genetic parameters and trends of the thigh muscle volume in Pannon White rabbits. In Proceeding of 9th World Rabbit Congress. 10-13 June. Verona, Italy. 2008. p. 115-119

[39] Niranjan SK, Sharma SR, Gowane GR. Estimates of Direct and Maternal Effects on Growth Traits in Angora Rabbits. Asian-Australasian Journal of Animal Sciences. 2010; 23(8): 981-986. DOI: 105713/ajas.2010.90549

[40] Peiró R, Badawy AY, Blasco A, Santacreu MA. Correlated responses on growth traits after two-stage selection for ovulation rate and litter size in rabbits. Animal. 2019;13(11):2457–2462. DOI:10.1017/S1751731119001423

[41] Piles M, Sánchez JP. Use of group records of feed intake to select for feed efficiency in rabbit. Journal of Animal Breeding and Genetic. 2019;136: 475-483. DOI: 10.1111/jbg.12395

[42] Su G, Kjaer JB, Brenøe UT, Sørensen P. Estimates of genetic parameters in Danish white rabbits using an animal model: I. Growth and carcass traits. World Rabbit Science. 1999;7:59-64. DOI: 10.4995/wrs.1999.381

[43] Sakthivel M, Balasubramayam D, Kumarasamy P, Gopin H, Raja A, Anilkumar R, Devaki A. Estimates of (co)variance components and genetic parameters for body weights and growth efficiency traits in the New Zealand white rabbit. World Rabbit Science. 2017;25(4):329-338. DOI: 10.4995/wrs.2017.7057

[44] Garreau H, David I, Hurtaud J, Drouilhet L, Gilbert H. Parameters for growth and feed efficiency traits in two commercial lines. In Proceeding of 10th World Congress on Genetics Applied to Livestock Production, 17-22 August, Vancouver, Canada. 2014. p. 565-567

[45] Larzul C, Rochambeau H. de. Selection for residual feed consumption

in the rabbit. Livestock Production Science. 2005;95(1-2):67-72. DOI: 10.1016/j.livprodsci.2004.12.007

[46] Blasco A, Nagy I, Hernández P. Genetics of growth and meat quality in rabbits. Meat Science. 2018;145: 178-185. DOI: 10.1016/j. meatsci.2018.06.030

[47] Johnson ZB, Harris DJ, Brown CJ, Robert WRG, Harrold RL. Genetic Analysis of Litter Size, Mortality and Growth Traits of New Zealand White Rabbits. The Professional Animal Scientist. 1988;4(2):11-16. DOI: 10.15232/S1080-7446(15)32340-8

[48] Rochambeau H de, Bolet G, Tudela F. Long-term selection. Comparison of two rabbit strains. In Proceeding of 5th World Congress on Genetics Applied to Livestock Production, 7-12 August, Guelph, Canada. 1994. p. 257–260

[49] Bole G. Problèmes liés à l'accroissement de la productivité chez la lapine reproductrice. INRA. Productions Animales 1998; 11(3): 235-238

[50] Argente MJ. Major components in limiting litter size. In: Payan-Carreira R, editor. Insights from animal reproduction. London, UK: InTech; 2016. p. 87-114. DOI: 10.5772/62280

[51] Gómez EA, Baselga M, Rafel O, García ML, Ramón J. Selection, diffusion and performances of six Spanish lines of meat rabbit. In Testik A, Baselga M. editors. 2. International Conference on Rabbit Production in Hot Climates. Zaragoza, Spain: CIHEAM, 1999. p. 147-152

[52] Garreau H, Piles M, Larzul C, Baselga M, Rochambeau H. de. Selection of maternal lines: last results and prospects. In Proceeding of 8th World Rabbit Congress. 7-10 September, Puebla, Mexico. 2004. p. 4-25

[53] Bolet G, Saleil G. Strain INRA2066 (France). Khalil MH, Baselga M. editors. In Rabbit genetic resources in Mediterranean countries. Zaragoza: CIHEAM, (Options Méditerranéennes: Série B. Etudes et Recherches; n. 38). 2002. p. 121-124

[54] Bolet G, Saleil G. Strain INRA2666 (France). Khalil MH, Baselga M. editors. In Rabbit genetic resources in Mediterranean countries. Zaragoza: CIHEAM, (Options Méditerranéennes: Série B. Etudes et Recherches; n. 38). 2002. p. 135-137

[55] Bolet G, Saleil G. Strain INRA9077 (France). Khalil MH, Baselga M. editors. In Rabbit genetic resources in Mediterranean countries. Zaragoza: CIHEAM, (Options Méditerranéennes: Série B. Etudes et Recherches; n. 38). 2002. p. 129-131

[56] Garreau H, Rochambeau H de. La sélection des qualités maternelles pour la croissance du laperau. In Proceeding of 10èmes Journées Recherche Cunicole, November, Paris, France. 2003. p. 61-64

[57] Baselga M, Ragab M, Mínguez C, El Nagar A.G. Analysis of methods to found new rabbit lines. Egyptian Journal of Rabbit Science. 2017;27(2):155-169.

[58] Gómez EA, Rafel O, Ramón J. The Prat Strain (Spain). Khalil MH, Baselga M. editors. In Rabbit genetic resources in Mediterranean countries. Zaragoza: CIHEAM, (Options Méditerranéennes: Série B. Etudes et Recherches; n. 38). 2002. p. 203-208

[59] Matics ZS, Nagy I, Gerencsér ZS, Randai I, Gyovai P, Donkó T, Dalle Zotte A, Curik I, Szendro ZS. Pannon breeding program in rabbit at Kaposvár University. World Rabbit Science. 2014; 22:287-300. DOI:10.4995/wrs.2014.1511

[60] Abou Khadiga GY, Youssef MK, Baselga M. Characterization of reproductive performance of the APRI

line of rabbits. In Proceeding of 10th World Rabbit Congress, 3-6 September. Sharm El-Sheikh, Egypt. 2012. p. 743-747

[61] Bolet G, Zerrouki N, Gacem M, Brun JM, Lebas F. Genetic parameters and trends for litter and growth traits in a synthetic line of rabbits. In proceedings 10th World Rabbit Congress, 3-6 September, Sharm El-Sheikh, Egypt. 2012. p. 195-199

[62] Capra G, Blumetto O, Elizalde E. Meat rabbit production in Uruguay. In Proceeding of 7th World Rabbit Congress, 7-10 July, Valencia, Spain. 2000. p. 51-58

[63] Gómez EA, Rafel O, Ramón J. The Caldes Strain (Spain). Khalil MH, Baselga M. editors. In Rabbit genetic resources in Mediterranean countries. Zaragoza: CIHEAM, (Options Méditerranéennes: Série B. Etudes et Recherches; n. 38). 2002. p. 193-198

[64] Bolet G, Saleil G. Strain INRA1077 (France). Khalil MH, Baselga M. editors. In Rabbit genetic resources in Mediterranean countries. Zaragoza: CIHEAM, (Options Méditerranéennes: Série B. Etudes et Recherches; n. 38). 2002. p. 113-116

[65] Lopéz M, Sierra I. The Gigante de España Breed (Spain). Khalil MH, Baselga M. editors. In Rabbit genetic resources in Mediterranean countries. Zaragoza: CIHEAM, (Options Méditerranéennes: Série B. Etudes et Recherches; n. 38). 2002. p. 213-220

[66] Randi E, Scossiroli RE. Genetic analysis of production traits in Italian new Zealand White and California pure-bred population. Genet Selection Evolution. 1980;12:296. DOI: 10.1186/1297-9686-12-3-296

[67] Argente MJ, Santacreu MA, Climent A, Blasco A. Relationships between uterine and fetal traits in rabbits selected on uterine capacity. Journal of Animal Science. 2003; 81(5): 1265-1273. DOI: 10.2527/2003.815265x

[68] Laborda P, Santacreu MA, Blasco A, Mocé ML. Selection for ovulation rate in rabbits: direct and correlated responses estimated with a cryopreserved control population. Journal of Animal Science. 2012;90(10):3392–3397. DOI:10.2527/jas.2011-4837

[69] Blasco A, Ortega JA, Climent A, Santacreu MA. Divergent selection for uterine capacity in rabbits. I. Genetic parameters and response to selection. Journal of Animal Science. 2005;83(10): 2297–2302. DOI: 10.2527/2005.8310229x

[70] Bolet G, Santacreu MA, Argente MA, Climent A, Blasco A. Divergent selection for uterine efficiency in unilaterally ovariectomized rabbits. I. Phenotypic and genetic parameters. In Proceeding 5th World Congress on Genetics Applied to Livestock Production, Guelph, Canada. 1994.p. 261-264

[71] Badawy AY, Peiró R, Blasco A, Santacreu, MA. Correlated responses on litter traits and survival traits after two-stage selection for ovulation rate and litter size in rabbits. Animal. 2019; 13: 453–459. DOI: 10.1017S175173111800233

[72] Argente MJ, Santacreu MA, Climent A, Bolet G, Blasco A. Divergent selection for uterine capacity in rabbits. Journal of Animal Science. 1997;75(9):2350-2354. doi:10.2527/1997.7592350x

[73] Santacreu MA, Mocé ML, Climent A, Blasco A. Divergent selection for uterine capacity in rabbits. II. Correlated response in litter size and its components estimated with a cryopreserved control population. Journal of Animal Science. 2005;83(10): 2303-2307. doi:10.2527/2005.83102303x

[74] Sánchez JP, Theilgaard P, Mínguez C. Baselga M. Constitution and evolution of a long-lived productive rabbit line. Journal of Animal Science. 2008;86:515-525. DOI: 10.2527/jas.2007-0217

[75] Larzul C, Ducrocq V, Tudela F, Juin H, Garreau H. The length of productive life can be modified through selection: an experimental demonstration in the rabbit. Journal of Animal Science. 2014;92: 2395-2401. DOI:10.2527/jas.2013-7216

[76] Moura A, Kaps M, Vogt DW, Lamberson WR. Two-way selection for daily gain and feed conversion in a composite rabbit population. Journal of Animal Science. 1997;75:2344–2349. DOI:10.2527/1997.7592344x

[77] Drouilhet L, Achard CS, Zemb O, Molette C, Gidenne T, Larzul C, Rousche J, Tircazes A, Segura M, Bouchez T, Theau-Clément M, Joly T, Balmisse E, Garreau H, Gilbert H. Direct and correlated responses to selection in two lines of rabbits selected for feed efficiency under ad libitum and restricted feeding: I. Production traits and gut microbiota characteristics. Journal of Animal Science. 2016;94,38–48. DOI: 10.2527/jas.2015-9402

[78] Garreau H, Gilbert H, Molette C, Larzul C, Balmisse E, Ruesche J, Secula-Tircazes A, Gidenne T, Drouilhet, L. Réponses à la sélection pour deux critères d'efficacité alimentaire chez le lapin. 1. Croissance, ingéré et efficacité alimentaire. In Proceeding of 16èmes Journées de la Recherche Cunicole, Le Mans, France. 2015. p. 161–164.

[79] Garreau H, Gilbert H, Molette C, Larzul C, Balmisse E, Ruesche J, Secula-Tircazes A. Direct and correlated responses to selection in two lines of rabbits selected for feed efficiency under ad libitum and restricted feeding. In Proceedings of 11th World Rabbit Congress, Qingdao. China. 2016. p. 43-4

[80] Zomeño C, Hernández P, Blasco, A. Divergent selection for intramuscular fat content in rabbits. I. Direct response to selection. Journal of Animal Science. 2013;91:4526–4531. DOI: 10.2527/jas2013-6361

[81] Martínez-Álvaro M, Hernández P, Blasco, A. Divergent selection on intramuscular fat in rabbits: Responses to selection and genetic parameters. Journal of Animal Science. 2016; 94:4993–5003. DOI: 10.2527/jas.2016-0590

[82] Martínez-Álvaro M, Penalba V, Hernández P, Blasco, A. Effect of divergent selection for intramuscular fat on sensory traits and instrumental texture in rabbit meat. Journal of Animal Science. 2016;94:5137–5143. DOI: 10.2527/jas.2016-0850

[83] Martínez-Álvaro M, Blasco A, Agha S, Hernández P. Correlated responses to selection for intramuscular fat in several muscles in rabbits. Meat Science. 2018; 139:187-192. DOI: 10.1016/j.meatsci.2018.01.026

[84] Hernández P, Aliaga S, Pla M, Blasco A. The effect of selection for growth rate and slaughter age on carcass composition and meat quality traits in rabbits. Journal of Animal Science. 2004;82:3138–3143. DOI: 10.2527/2004.82113138x

[85] Bodin L, Garcia M, Saleil G, Bolet G, Garreau H. Results of 10 generations of canalising selection for rabbit birth weight. In Proceeding of 9th World Congress on Genetics Applied to Livestock Production, August, Leipzig, Germany. 2010. p. 0391

[86] Garreau H, Bolet G, Larzul C, Robert-Granié C, Saleil G, San Cristobal M, Bodin L. Results of four generations of a canalising selection for rabbit birth weight. Livestock Science. 2008;119:55-62. DOI: 10.1016/j.livsci.2008.02.009

[87] Bolet G, Garreau H, Joly T, Theau-Clement M, Falieres J, Hurtaud J, Bodin L. Genetic homogenisation of birth weight in rabbits: Indirect selection response for uterine horn characteristics. Livestock Science. 2007; 111:28-32. DOI: 10.1016/j.livsci.2006.11.012

[88] Blasco A, Martínez-Álvaro M, García ML, Ibáñez-Escriche N, Argente MJ. Selection for genetic environmental sensitivity of litter size in rabbits. Genetic Selection Evolution. 2017; 49:48-55. DOI: 10.1186/s12711-017-0323-4

[89] Argente MJ, García ML, Zbyňovská K, Petruška P, Capcarová M, Blasco A. Correlated response to selection for litter size environmental variability in rabbits´ resilience. Animal. 2019;13:2348-2355 DOI: 10.1017/S1751731119000302

[90] Argente MJ, Calle EW, García ML Blasco A. Correlated response in litter size components in rabbits selected for litter size variability. Journal of Animal Breeding and Genetic. 2017;134:505-511 DOI: 10.1111/jbg.12283

[91] García ML, Blasco A, Argente MJ. Embryologic changes in rabbit lines selected for litter size variability. Theriogenology. 2016;86:1247-1250 DOI: 10.1016/j.theriogenology.2016.04.065

[92] Calle EW, García ML, Blasco A, Argente MJ. Correlated response in early embryonic development in rabbits selected for litter size variability. World Rabbit Science. 2017;25:323-327 DOI: 10.4995/wrs.2017.6340

[93] Garreau H, Brard S, Hurtaud J, Guitton E, Cauquil L, Licois D, Schwartz B, Combes S, Gidenne T. Divergent selection for digestive disorders in two commercial *Echerichia coli* 0-103. In Proceeding of 10th World Rabbit Congress, September, Sharm El-Sheikh, Egypt. 2012. p. 153-157

Chapter 6

# Profitability in Rabbit Breeding

*Mariam Pascual and Ernesto A. Gómez*

## Abstract

Rabbit production must be sustainable. One of the main bases of this sustainability is the profitability of the farms, in which improvement is achieved through the technical and economic management. These managements imply collecting production and economic data, calculating indexes, comparing results with those obtained in other farms, making decisions, and evaluating the consequences of changes. The chapter details the different steps to follow to develop the technical and economic management of the rabbit farms to improve their profitability. Templates to collect data, formulae to obtain different technical and economic indexes, results obtained in different producing countries, and possible techniques to improve profitability are shown.

**Keywords:** economic management, production costs, profitability, rabbit, technical management

## 1. Introduction

The order Lagomorpha includes the family Leporidae (rabbits and hares) and Ochotonidae (pikas). The species from the order with the highest interest in farming is the rabbit, with a production of almost 1.5 millions of tonnes of rabbit meat in 2018, which supposes 1.15% of the world pig meat production [1]. Hare is scarcely harvested in some countries, and its industrial production is marginal compared to rabbit meat production, and the breeding of pikas for meat consumption has not been developed. In consequence, the chapter will be focused on rabbit farming.

The main goal in rabbit meat production is to maximize the profitability respecting animal welfare and environmental concerns. The increase of the profitability of the farms is based on the knowledge of the technical and economic results of the farm, which will allow the detection of weak points that should be amended. The chapter will be focused on the procedure to develop technical and economic management in rabbit farms.

Technical and economic management in animal production implies (**Figure 1**):

- Technical and economic data collection
- Calculation of the technical and economic indexes
- Benchmarking: comparison of results obtained with those observed in other farms
- Detection of weak points that could be improved

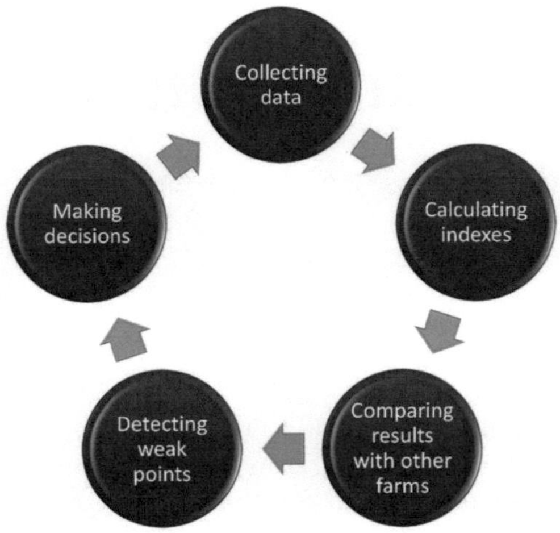


**Figure 1.**
*Technical and economic management continuous process.*

- Decision making to overcome the weak points
- Return to the first point for evaluating the effects of the changes applied

These points will be further discussed in the text.

It is essential to remember that technical and economic management are interrelated, and they are continually feeding back; therefore, both management processes have to be developed in the farm, because:

- Best technical indexes do not necessarily imply, surprisingly, the best economic results. Example: the increase in production and incomes obtained when increasing the number of workers might be lower than the increase in labor costs, leading to better technical but worse economic results.
- Economic management will indicate the profitability of the farm, but its improvement will be mainly achieved by the improvement of the technical indexes.

## 2. Technical and economic management

### 2.1 Data collection

#### *2.1.1 Technical data*

Rabbit females have their first insemination or mating (hereinafter, insemination) at approximately 4.5–5 months old. During their reproductive life, females are usually grouped in batches. Females from the same batch undergo the different events at the same time. That is, all the females belonging to the same batch are inseminated the same day; therefore kindling, weaning, and sale to the slaughterhouse after the fattening period of the young rabbits are achieved at similar dates for all the females in the batch. Moreover, the females of the farm can be all grouped in a single batch or in more than one batch, being displaced in time. **Table 1** shows the number of batches depending on the number of days between

| Interval length between delivering and insemination (cycle length) (days) | Interval length between consecutive inseminations (days) | | | | | | | | |
|---|---|---|---|---|---|---|---|---|---|
| | **7** | **14** | **21** | **28** | **35** | **42** | **49** | **56** | **63** |
| Postpartum (31–34)[1] | 5 | | | | 1 | | | | |
| 11 (42) | 6 | 3 | 2[2] | | | 1[2] | | | |
| 18 (49) | 7 | | | | | | 1 | | |
| 25 (56) | 8 | 4 | | 2 | | | | 1 | |
| 32 (63)[3] | 9 | | 3 | | | | | | 1 |

[1]*Intensive rhythm; not recommended*
[2]*The most frequent ones*
[3]*Extensive rhythm*

**Table 1.**
*Number of possible batches in the farm depending on the interval length between delivering and insemination and interval length between consecutive inseminations in the farm.*

delivering and insemination and number of days between two consecutive inseminations in the farm. For example, when females are inseminated 11 days postpartum, 1, 2, 3, or 6 batches in the farm are possible, and the number of days between two consecutive inseminations will be 42, 21, 14, or 7 days, respectively. When females are distributed in more than one batch, the batches can be treated as single independent batches (the female is assigned to one batch along its whole reproductive life) or multiple batches (the female changes the batch after a negative palpation to shorten the unproductive periods).

The technical data is usually collected in the farm by filling up a formulary as the different events of the batch occur. **Table 2** shows an example of formulary. The

| Date of insemination: 14-06-2019[1] | Data simulation[1,2] |
|---|---|
| Number of females[3] | 750 |
| Number of inseminations | 748 |
| Number of positive palpations | 592 |
| Number of aborts | 19 |
| Number of kindlings | 587 |
| Number of total kits born | 5870 |
| Number of kits born alive | 5518 |
| Number of weaned kits | 4861 |
| Number of rabbits produced | 4516 |
| Number of dead females | 43 |
| Number of culled females | 55 |
| Number of nulliparous | 104 |
| Kg rabbit produced | 9935 |
| Total feed consumed (kg)[4] | 32,175 |

[1]*Simulation of data. Adapted from [2].*
[2]*Artificial insemination at 11 days postpartum, single batch, weaning at 35 days old*
[3]*Rabbit females in the batch between first insemination and culling*
[4]*Total feed consumed in the farm, including maternity, replacement, lactating, fattening, males, etc.*

**Table 2.**
*Template for technical data collection in each batch of the rabbitry.*

table shows also a simulation of technical data. Later on, the data will be used to calculate the technical indexes by using different programs that are explained in the next subsection.

It is important to realize that filling one formulary for each batch in the farm is recommended. Collection of data could be simple in farms with a single batch, but it could be more complicated when the farm works with more than one. In those cases, the formulary could collect not the information from one batch but all the data produced in the farm in a fixed period of time. This practice is not recommended for different reasons. When this period of time is short (e.g., 1 month), the indexes calculated induce confusion, since the results obtained for the different events do not correspond to the same group of animals. For example, the number of produced kits per kindling in May will not correspond to the fertility calculated in May, as the inseminations corresponding to the animals produced in May occurred in February. If the period of time is longer (e.g., 6 months), the quality of the calculated indexes improves, but without reaching a full agreement, and, in addition, a huge delay is generated between the early events and the calculation of the indexes and the possible decision-makings. Moreover, problems in one batch could be hidden under the overall results. Finally, even when troubles are detected, there is no possibility of identifying which was the problematic batch.

### *2.1.2 Economic data*

There are many nomenclatures to define the breakdown of production costs. One of them classifies production costs as variable or as fixed ones. The different items in each group and a simulation of economic data can be seen in the example of formulary shown in **Table 3**. This classification is based on whether the cost would vary depending on the level of production, given farm size. In this way, variable costs will depend directly on the level of production. For example, the cost of feed during the fattening period will depend on the number of kits weaned in each batch. On the other hand, fixed costs do not depend on the production. For instance, expenses in telephone and the Internet are supposed to be independent on the level of production. However, it is sometimes difficult to classify some costs according to this criterion.

Recommendations on the length of the period of time for collecting economic data are similar to those for technical management. Collection and calculation per batch are recommended. Nevertheless, in practice, rabbit breeders often match the period considered with the dates established for legal and administrative obligations, recording on the same formulary the data generated every 3 months or even for a year.

There are three fixed costs that deserve a special mention, as they are not always considered when the economic management of the farm is applied. One of the most neglected items is the own (farmer) labor cost. Some farmers do not consider this labor as a cost and get their own earnings from income minus total costs (excluding salary, amortization cost and cost of capital). This practice should be avoided, as it could generate a false appearance of profitability.

Another problematic cost is the amortization because it is an intangible concept and not a monetary payment. This item represents the depreciation of the farm (facilities, machinery, and systems) in a period of time and has to be considered along the whole lifetime of the farm. Many times, the farm is constructed with capital obtained with a bank loan, and the amortization cost considered by the farmer is the capital returned to the bank in the period of time under study. That is, if the farm has a lifetime of 30 years, but the loan has to be returned in a maximum period of eight years, the farmer considers that there is an amortization cost equivalent to the loan repayments during 8 years and null amortization costs during the

| Period of time: 01-01-2019 to 31-12-2019[1] | | | |
|---|---|---|---|
| **Variable costs (€)** | | **Fixed costs (€)** | |
| Maternity feed | 21,900 | Own labor | 19,200 |
| Fattening feed | 45,375 | Own social security | 4100 |
| Replacement feed | 2625 | External labor | 3775 |
| Other feeds | 0 | External social security | 894 |
| Health | 10,725 | Manure management | 590 |
| Artificial insemination | 6518 | Energy and communications [2] | 8500 |
| Animals for replacement | 8850 | Corpse removal | 700 |
| Other variable expenses | 0 | Maintenance and repairs | 740 |
| | | Expendable material | 160 |
| | | Insurances and taxes | 2725 |
| | | Financial and banking | 0 |
| | | Credits and loans | 0 |
| | | Rental agreements | 0 |
| | | Other fixed expenses | 0 |
| | | Amortization plus cost of capital | 17,529 |
| **Incomes** | | | |
| Sales to slaughterhouse | 154,590 | | |
| Sales of alive animals | 1150 | | |
| Subsidies | 560 | | |
| Other incomes | 0 | | |
| Inventory differences[3] | 100 | | |

[1]*Simulation of data. Adapted from [2]*
[2]*Water, power, gas, fuel, telephone, Internet, nest material*
[3]*Feed, animals, etc.*

**Table 3.**
*Template for cost data collection in the rabbitries.*

rest of the lifetime of the farm. However, this approach should be avoided, since this will lead to underestimations of the profitability at the beginning of the activity and overestimations at the last years. A more appropriated solution is calculating the real quantity of capital invested in the farm and distributing the cost along the whole lifetime of the farm, for instance, the construction of a farm that involved an investment of 300,000 euros. In the case where the capital was obtained through a business loan and should be returned in 8 years, the total capital investment increases, not only for the interests of the loan but also because of the devaluation of the money (one euro in 2020 will have a higher value than one euro in 2028 in inflation conditions). Therefore, capital investment should be corrected before calculating the cost per year. For example, considering a loan interest of 2.5%, the total capital would increase around 333,200 euros, and annual inflation of 1% for the next 8 years, the total capital would be established in around 361,800 euros. That is, the amortization of the farm will be 12,060 euros per year, but please note that amortization represents the devaluation of the facilities and material, which decreases with time; thus not constant but decreasing amortization must be calculated for each year.

Finally, the most controversial cost is the cost of capital. This intangible cost represents the minimal return that the farmer would expect if he made a capital investment in that farm and not in another investment. For example, the farmer might not invest that capital of money in a farm unless the profit was three times higher than the profit obtained when the capital was invested in a fixed-term deposit. This cost is hardly calculated and considered in the economic management of the farm. However, its mention is mandatory, because its estimation is compulsory not only in the economic management but also before making any investment, for example, in a rabbit farm. As an example, considering 0.75% of annual percentage rate (APR), 300,000 euros would rent 1823 euros in 1 year (after taxes); thus, the farmer could consider the investment in a farm only if a profit of at least three times higher (5469 euros) was expected.

## 2.2 Calculation of the technical and economic indexes

Once data has been collected in templates similar to those shown in **Tables 2** and **3**, data can be registered in (a) spreadsheets, which will allow the calculation of the results in the farm, or (b) management programs that will allow both the calculation and the comparison with other farms. This section will be focused on the first option, and the second one will be developed in Section 3.

### *2.2.1 Technical indexes*

The main technical indexes calculated in rabbitries are shown in **Table 4**. Indexes related to fertility (apparent fertility, real fertility, and kindling interval) reflect the success in the different litter size components: ovulation rate, fertilization rate, and prenatal survival. It is important to notice that:

- The apparent fertility is calculated with the results of the test of pregnancy of the females, which consists in an abdominal palpation 10–14 days after insemination. When the female is mated or inseminated, the ovula in the follicles are liberated, and the follicles become corpus luteum, which liberate progesterone to maintain the possible pregnancy and prevent new ovulations. In the case of negative pregnancy, at 17 days of mating, the uterus starts the liberation of prostaglandins that eliminate the corpus luteum [3]. Therefore, females with negative palpation can be inseminated again in the next batch as long as inseminations differ in more than 17 days, although the period can be shortened with the inoculation of prostaglandins. In farms with one single batch, the early diagnosis of pregnancy is useless, and palpation is usually developed around 25 days post-insemination to plan the nest arrangement.

- The indexes related to fertility can be calculated per female, as shown in **Table 4**, or per insemination, replacing in the formulae the number of females by the number of inseminations in the denominator. Therefore, it is imperative to compare the results only with farms that used the same formulae for the calculation of the index.

- The kindling interval depends both on the cycle length and real fertility. In farms with a single batch or single independent batches (see Section 2.1.1), the index can be calculated with the data of the batch. However, in farms with multiple batches, females with negative palpation are changed to other batches; therefore the index must be calculated considering the real fertility achieved in all the batches as a whole.

| Indexes and formulae | Data simulation[1] |
|---|---|
| Apparent fertility (%)[2] | |
| $AF\ (\%) = \frac{\text{number of positive palpations}}{\text{number of females}}\ x\ 100$ | $AF(\%) = \frac{592}{750}\ x\ 100 = 78.9\%$ |
| Real fertility (%)[2] | |
| $RF\ (\%) = \frac{\text{number of kindlings}}{\text{number of females}}\ x\ 100$ | $RF\ (\%) = \frac{587}{750}\ x\ 100 = 78.3\%$ |
| Cycle length (days) | |
| $CL = \text{lenght of gestation (days)}$ $+\text{interval kindling to insemination (days)}$ | $CL = 31 + 11 = 42\ \text{days}$ |
| Number of batches per year | |
| $B = \frac{365}{CL}$ | $B = \frac{365}{42} = 8.69\ \text{batches}$ |
| Kindling interval (days)[3] | |
| $KI = \frac{CL}{\left(\frac{RF}{100}\right)}$ | $KI = \frac{42}{0.783} = 53.7\ \text{days}$ |
| Abortions (%) | |
| $A\ (\%) = \frac{\text{number of abortions}}{\text{number of inseminations}} \times 100$ | $A\ (\%) = \frac{19}{748} \times 100 = 2.5\%$ |
| Mortinatality (%) | |
| $BM(\%) = \frac{(\text{total kits born} - \text{kits born alive})}{\text{total kits born}} \times 100$ | $BM\ (\%) = \frac{(5870-5518)}{5870} \times 100 = 6.0\%$ |
| Lactation mortality (%) | |
| $LM\ (\%) = \frac{(\text{kits born alive} - \text{weaned kits})}{\text{kits born alive}} \times 100$ | $LM\ (\%) = \frac{(5518-4861)}{5518} \times 100 = 11.9\%$ |
| Fattening mortality (%) | |
| $FM\ (\%) = \frac{(\text{weaned kits} - \text{rabbits produced})}{\text{weaned kits}} \times 100$ | $FM\ (\%) = \frac{(4861-4516)}{4861} \times 100 = 7.1\%$ |
| Kits born alive per kindling[4] | |
| $BAK = \frac{\text{number of kits born alive}}{\text{number of kindlings}}$ | $BAK = \frac{5518}{587} = 9.4$ |
| Kits weaned per kindling[4] | |
| $WK = \frac{\text{number of weaned kits}}{\text{number of kindlings}}$ | $WK = \frac{4861}{587} = 8.3$ |
| Rabbits produced per kindling[4] | |
| $PRK = \frac{\text{number of rabbits produced}}{\text{number of kindlings}}$ | $PRK = \frac{4516}{587} = 7.7$ |
| kg rabbit produced per kindling[4] | |
| $kgPRK = \frac{\text{kg rabbits produced}}{\text{number of kindlings}}$ | $kgPRK = \frac{9935}{587} = 16.9$ |
| Liveweight for sale (kg) | |
| $LW\ (kg) = \frac{\text{kg rabbit produced}}{\text{number of rabbits produced}}$ | $LW\ (kg) = \frac{9935}{4516} = 2.2$ |
| Global feed conversion rate (kg/kg) | |
| $GFCR = \frac{\text{kg consumed feed}}{\text{kg rabbit produced}}$ | $GFCR = \frac{32175}{9935} = 3.2\ kg/kg$ |
| Dead females (%) | |
| $DF\ (\%) = \frac{\text{number of dead females}}{\text{number of females}} \times 100$ | $DF\ (\%) = \frac{43}{750} \times 100 = 5.7\%$ |
| Culled females (%) | |
| $CF\ (\%) = \frac{\text{number of culled females}}{\text{number of females}} \times 100$ | $CF\ (\%) = \frac{55}{750} \times 100 = 7.3\%$ |
| Nulliparous females (%) | |
| $NF\ (\%) = \frac{\text{number of nulliparous}}{\text{number of females}} \times 100$ | $NF\ (\%) = \frac{104}{750} \times 100 = 13.9\%$ |

| Indexes and formulae | Data simulation[1] |
|---|---|
| Kindlings per female and year | |
| $KFY = \frac{365}{KI}$ | $KFY = \frac{365}{53.7} = 6.8$ |
| Kits born alive per female and year | |
| $BAFY = KFY \times BAK$ | $BAFY = 6.8 \times 9.4 = 63.8$ |
| Kits weaned per female and year | |
| $WFY = KFY \times WK$ | $WFY = 6.8 \times 8.3 = 56.3$ |
| Produced per female and year | |
| $PRFY = KFY \times PRK$ | $PRFY = 6.8 \times 7.7 = 52.3$ |
| kg produced per female and year (kg) | |
| $kgPRFY = PRFY \times LW$ | $kgPRFY = 52.3 \times 2.2 = 115.0$ kg |
| Replacement rate (%) | |
| $RR (\%) = B \times NF$ | $RR (\%) = 8.69 \times 13.9\% = 120.8\%$ |

[1]*Simulation of results calculated with data from* ***Table 2***
[2]*The index is sometimes calculated as the ratio with the number of females rather than the number of inseminations.*
[3]*In farms with multiple batches, the term RF should be replaced by ($\sum$(kindlings in n batches))/number of females in the farm), where n is the number of batches in the farm.*
[4]*The index can be calculated per female or by insemination by replacing "number of kindlings" in the formulae by "number of females" or "number of inseminations," respectively.*

**Table 4.**
*Formulae of the main technical indexes calculated in technical management of rabbit farms.*

Indexes for mortality of the offspring will be mainly influenced by the sanitary status of the herd, but it is important to remember that these indexes are also highly dependent on the moment the data is collected. In this way, the number of stillborn will be higher if data is not collected just after birth but some hours later, and the number of weaned and produced young rabbits is conditioned to the age at weaning and sale, respectively. Once more, results obtained must be compared with farms with similar management.

In the same way, the number of kits born alive, kits weaned, rabbits produced, and kg of rabbit produced will also depend on the moment of data collection. These data are used to analyze the productivity of the batch per female, per insemination, or per kindling. The indexes per kindling will reflect the success in prolificacy and survival, but the indexes per insemination or female will include also the success in fertilization. Usually, markets demand a fixed weight at slaughter; however, in markets where the produced animals are sold at a fixed age, indexes obtained for kg of rabbit produced will reflect also the success in daily weight gain.

Another summary index, more modern and with a lot of popularity, is the number of kg of rabbit sold per insemination carried out. It includes the successes obtained in both the reproduction and in the fattening sections but is easily increased by slaughtering at higher ages when the market allows heavier carcasses. In Galicia (Spain), 24% of the farms under management produces more than seven rabbits per insemination [4]. In France, the coefficient of variation of the index kg produced per insemination is around 18% [5].

The global feed conversion rate indicates the kg of feed needed to produce a kg of rabbit. The feed considered for its estimation is the total feed consumed by all the animals in the farm, as reproductive females, replacement, fattening, males, etc.

Finally, the productivity of the batch can be also expressed by female and year by extrapolating the information of the batch to the whole year. The indexes will be obtained by multiplying the indexes per female by the number of batches per year (B) or by multiplying the indexes per kindling by the number of kindlings per female and year (KFY).

### *2.2.2 Economic indexes*

It is not enough to know how much money goes in and out. The economic indexes will show the real situation of the profitability of the farm. There are several indexes that are calculated to express the economic situation of the farm.

Incomes include the sale of animals to the slaughterhouse, the possible subsidies received by the activity, and the inventories differences (positive or negative) of the value of the breeders, and the stored feed must also be computed.

One of the most widely used indexes for its simplicity and ease of calculation is the income over feed cost margin (IFCM), which is usually calculated as the difference between incomes and feed cost. This index is frequently calculated and shown in reports of technical and economic management of rabbitries because of two main reasons. First, the index reflects the margin of profitability left after considering the feeding cost, which is the greatest production cost in the rabbitries [2]. Second, its calculation is not complicated because, in general, it only requires operation with the invoices from the slaughterhouse and the feed company.

The economic analysis of the farm must go further than the calculation of IFCM, as there is still a huge range between the index and the real profitability of the farm. From an economic point of view, there are different terminologies to refer to the margins used for benchmarking. An index more accurate is the gross margin, estimated as the difference between incomes and variable costs. These variable costs include not only the feed but mainly health, artificial insemination, and animals for replacement expenses, which should be also reflected in invoices. Therefore, its estimation is also feasible. It does not measure the profit of the enterprise. It shows how well sales cover the direct costs related to the production of goods (especially expressed in percentage of incomes).

The real profitability of the farms is finally obtained calculating the net margin, as the difference between incomes and total costs. This is the quantity earned by the business owner. Net margin is an indicator of the result of the exploitation that must serve to sustain the own salary and remunerate the invested capital. Net margin (without subsidies) is related to farm viability. It shows how profitable the company is when compared to its past self or to other farms, expressed as a rate (by kg of rabbit sold, by number of females, by number of inseminations, etc.). Net margin reflects the company's ability to generate profit for owners. If the farmer cannot save, a situation of decapitalization will occur. It is important to highlight that one of the fixed costs is the salary of the farmer (own labor). Generally, the farmer is also the business owner, getting both the net margin plus the own salary. This profitability is known as family net margin.

These economic indexes are frequently shown not per period of time but per female and year or per kg of rabbit produced. Formulae in **Table 5** correspond to the calculation of the indexes for a given period of time. Data used in the simulation are shown in **Table 3** and correspond to a period of 1 year. The net margin per year and female must be calculated as 1494€/(1year × 750 females), that is, the net margin is 1.99€ per female and year. In the same way, the net margin per kg produced may be obtained as the net margin divided by the number of kg produced in that period of time. For a total production of, e.g., 86,340 kg in 1 year, the net margin will be 1494€/86,340 kg, that is, 0.017€ per kg of rabbit produced.

| **Indexes and formulae (€)**[1] | **Data simulation**[2] |
|---|---|
| Variable costs | |
| VC = ∑ variable costs | VC = 21, 900 + 45, 375 + 2, 625 + 0 + 10, 725 + 6, 518 +8, 850 + 0 = 95, 993€ |
| Fixed costs | |
| FC = ∑ fixed costs | FC = 19, 200 + 4, 100 + 3, 775 + 894 + 590 + 8, 500 + 700 +740 + 160 + 2, 725 + 17, 529 = 58, 913€ |
| Production cost | |
| PC = VC + FC | PC = 95, 993 + 58, 913 = 154, 906€ |
| Incomes | |
| I = ∑ incomes | I = 154, 590 + 1, 150 + 560 + 0 + 100 = 156, 400€ |
| Income over feed cost margin | |
| IFCM = incomes − feed | IFCM = 156, 400 − (21, 900 + 45, 375 + 2, 625 + 0) = 86, 500€ |
| Gross margin | |
| GM = I − VC | GM = 156, 400 − 95, 993 = 60, 407€ |
| Net margin | |
| NM = I − (VC + FC) | NM = 156, 400 − (95, 993 + 58, 913) = 1, 494€ |
| Family net margin | |
| FNM = I − (VC + FC) + own labor | FNM = 156, 400 − (95, 993 + 58, 913) + 19, 200 = 20, 694€ |

[1]*All the indexes are also frequently calculated: (a) per female and year by dividing the value obtained along 1 year by the number the females in the farm; (b) per kg rabbit produced by dividing the value between the number of kg produced in the period of time considered.*
[2]*Simulation of results calculated with data from **Table 3***

**Table 5.**
*Formulae of the main economic indexes calculated in economic management of rabbit farms.*

## 3. Detecting weak points: index evolution and benchmarking

### 3.1 Evolution of the technical and economic indexes

The first tool to detect weak and strength points in a farm with technical and economic management is observing the evolution of the results along time, which will show changes between batches. These changes will be a consequence of variation in:

- External factors that cannot be controlled, as high variations in the climate conditions along seasons, market prices, etc.
- Factors that might be controlled, as problems in the insemination process, quality of the water, feed composition, etc.

And always bearing in mind that there is also a "natural" variation between batches: the same farmer, with a similar climate, diet, and husbandry, may have slight variations between the indexes of two twin rabbitries.

The main objective when analyzing the evolution of the indexes will be to distinguish if the factor affecting the results can be controlled to improve the situation.

### 3.2 The importance of comparing with others

The absence of variations in the technical and economic index evolution might not necessarily imply the inexistence of weak points that could be improved. These points can be detected by comparing the results with those obtained in other farms. This technique is called benchmarking. Benchmarks are reference points that you use to compare your performance against the performance of others. Benchmarking is commonly used to compare costs or technical and economic performances.

The first step for the comparison might be complicated as it implies access to the results from other farms:

- The technical and economic management is highly important in the rabbit farms but is less common than expected. Some reasons could be the lack of observing the advantages of the management at a short term and the false convincement of knowing the real situation of the farm just doing management in their head.

- Rabbit farmers doing management are sometimes reticent to show their indexes, probably due to fear to negative valuations that other people may make and/or the fear of promoting the competence among colleagues.

As commented along Subsection 2.2.1, the indexes obtained in the farm must be compared to the indexes obtained in farms with similar characteristics:

- Size: in large farms a part of the fixed costs is diluted, although it can strongly increase the labor costs.

- Market conditions: there are differences between countries in input prices and in the weight and price of live rabbits at the time of sale.

- Socioeconomic conditions: it makes no sense to compare an industrial farm with a backyard production.

- A similar management: for example, the number of total born per female and year must not be compared between two farms with 42 and 49 days of cycle length, as the number of cycles per year is higher in the former, and the mortality during lactation must not be compared between two farms weaning at different ages, as it is not possible to distinguish if differences are due to health problems or to the length of the lactation.

- A similar data recording and calculation process: for instance, some farms do not inseminate the females that do not show a favorable state of the vulva; therefore the indexes per insemination will be overestimated and cannot be compared with farms where all the females are inseminated regardless of the color of the vulva.

Obtaining technical and economic indexes from the farm will provide a valuable material to detect weak points and make decisions to improve the profitability of the farm. It is important to realize that the main and most important tool to analyze and make decisions is the specialists in cuniculture who are in direct contact with the farm, such as technicians and veterinarians.

In general, comparisons between farms with different management techniques should not be made. However, in some cases, these comparisons could be very useful comparing indexes that involve the information of total productions over long periods of time. For example, farmers with different lengths of the reproductive cycle should not compare the total born alive per female and year but the total number of fryer rabbits (or kg) produced per year.

If a farmer detects that one or more of his/her technical indexes are worse than in other similar farms, it is mandatory to be able to discover the possible causes, which can be multiple: they could only differ in one management technique, although it could also be due to a state of health different between herds or to variations in environmental control.

### 3.3 Programs allowing the comparison of results

The best option for technical and economic management is the use of collective management programs that allow the comparison of the results with those obtained by other rabbit breeders. Usually, the user has no access to the indexes obtained in one specific farm but to the mean obtained in a reference group, which is formed with an ensemble of rabbitries with similar characteristics to the farm under review. These platforms usually have the following characteristics:

- The farmer has access to templates for data collection that can be printed or filled up online.
- Indexes are calculated automatically by the platform.
- Results can be compared immediately with those obtained in reference groups.
- Frequently, the farmer has personal advice from the developer of the platform or associated technicians in the detection of weak points and the search for possible solutions.

These platforms are usually free and are developed and managed by:

- Agricultural technical institutes from governmental entities. Examples of these platforms in France are RENALAP and RENACEB, by the Aviculture Technical Institute (ITAVI). RENALAP was developed in 1983 for farms where not the batch but the individual data generated per female was recorded, and RENACEB was created in 1995 for farms with management in batches [6]. Other examples in Spain are the **bdcuni** (Rabbit Sector Database), which was developed in 2008 by the Valencian Institute of Agrarian Research (IVIA) [7] for farms with management in batches, and the management program developed by the Institute for Agrifood Technology and Infrastructures of Navarra (INTIA) [8].
- Companies that provide the service to the farmers buying their products or agricultural cooperatives which offer the platform to their partners.

### 3.4 Technical and economic indexes in different countries

Rabbit meat is mainly produced in China (865,477 tonnes), Spain (55,824 tonnes), France (43,886 tonnes), and Italy (43,109 tonnes) (data from 2018 [1]). Management and markets differ between countries, and this is reflected in the technical and economic performances obtained.

| | [4] | [8] | [9] |
|---|---|---|---|
| | Spain | Spain | France |
| Year | 2018 | 2018 | 2017 |
| No. of farms | 88 | 12 | 697 |
| No. of females | 57,816 | 11,307[1] | 454,444[1] |
| No. of females per farm | 657 | 942 | 652 |
| Replacement rate (%) | 125 | | |
| No. of nulliparous rate per batch (%) | | | 14.1 |
| No. of total born per kindling | | | 10.8 |
| No. of born alive per kindling | 10.90[2] | | 10.2 |
| Apparent fertility (%) | 85.2 | | |
| Real fertility (%) | 78.4 | | 82.5 |
| Mortinatality (%) | | | 5.6[1] |
| Lactation mortality (%) | 11.41 | | 15.6[1] |
| Fattening mortality (%) | 9.8 | | 8.7[1] |
| No. of weaned per kindling | 9.66[1] | | 8.60 |
| No. of weaned per insemination | 7.57[1] | | 7.13 |
| No. of produced per kindling | 8.71[1] | | 7.85 |
| No. of produced per insemination | 6.38 | | 6.48 |
| kg produced per insemination | 14.44 | | 16.01 |
| No. of produced per female and year | | 53.3 | 52.3 |
| Mean slaughter weight (kg) | 2.261[1] | 2.13 | 2.47 |
| Global feed conversion rate (g/g) | 3.52 | | 3.34 |
| Age at sale (days) | | | 73.4 |
| No. of kindlings per female and year | | | 6.66 |

[1]*Estimated*
[2]*Number of born alive and considered as viable*

**Table 6.**
*Mean technical indexes in some management programs.*

**Tables 6** and **7** include a relation of differences in technical and economic results in Spanish and French programs. In general, results shown belong to databases with farms working mainly with a cycle of 42 days. In general, rabbit breeding techniques are quite similar in France and Spain, but there are few differences in both management and markets that have to be considered:

- French consumers demand heavier carcass weight than Spanish ones; the higher slaughter weight in the French rabbitries supposes an increase of the profitability.

- Official data on technical management in the Spanish rabbit sector is lacking. Huge databases generated in Spanish farms are probably owned by private sector firms, and the accessible published data is scarce, and data shown in the text is data from regional associations. Meanwhile, France is collecting and reporting technical information from a large and therefore more representative

| | [8] | [9] | [2] |
|---|---|---|---|
| | Spain | France | Spain |
| | 2018 | 2017 | 2012 |
| | €/kg | €/kg | % total costs |
| **Incomes** | **1.834** | | |
| Sales of young rabbits to the slaughterhouse | 1.834 | 1.75 | |
| **Variable costs** | **1.191** | | |
| Feeding | 0.935 | 0.89 | 45.2 |
| Zoosanitary products | 0.160 | | 6.9 |
| Insemination, renewal, and other | 0.097 | | 9.9 |
| **Fixed costs** | **0.412** | | |
| Labor and social security (external and owner) | | | 18.1 |
| External labor and social security (external and owner) | 0.148 | | |
| Consumptions (water, power, gas, fuel, telephone, Internet, etc.) | 0.075 | | 7.0 |
| Administration | | | 3.1 |
| Amortization and cost of capital | | | 9.8 |
| Amortization and rental agreements | 0.100 | | |
| Others | 0.089 | | |
| **Total production costs** | **1.603** | | |
| **Margins** | | | |
| IFCM[2] | 0.899[1] | 0.92 | |
| IFCM[2] (€/doe and year) | 102.1[1] | 119.5 | |
| IFCM[2] (€/insemination) | | 14.84 | |
| Feed cost respect to kg rabbit sold | | 0.89 | |
| Feed cost respect to total production cost (%) | 58.3[1] | | |
| Feed cost respect to rabbit sale income (%) | | 50.8 | |
| Gross margin | 0.643[1] | | |
| Net margin | 0.231 | | |
| Net margin (€/rabbit) | 0.490 | | |

[1]*Estimated from other indexes*
[2]*Income over feed cost margin*

**Table 7.**
*Outputs of economic management programs.*

database through its program RENALAP and RENACEB [9]. The economic results are, however, scarce in both countries.

Comparison of results with those found in other countries has to be done prudently, as the management system in each market can clearly condition the values obtained. An example might be the case of Belgium, which has observed an increase of mortality during fattening with an associated decrease of the number of rabbits produced per female and year when changing from cages to park system [10].

## 4. Making decisions

### 4.1 The process of making decisions

Once the weak points are detected, the farmers and technician need to make decisions to increase profitability by the improvement of the technical indexes.

There are at least two ways to find out which techniques might lead to an improvement:

- The comparison with other farms with similar management will show which indexes should be improved, but the comparison with the means of an ensemble of farms that differ at least in one management characteristic might indicate an improvement in the results if the difference in management is applied in the farm.

- Previous studies may indicate how to improve some indexes. Some examples are detailed in the next section.

### 4.2 Possible management techniques to improve technical indexes

As mentioned, the main assistance should be the advice of the veterinarians and technicians. However, some examples of possible measures or recommendations to improve the technical indexes are detailed below:

Fertility: the index can be affected by (a) body condition (lower when the percentage of fat is too low or too high); (b) physiological status (lower fertility in lactating females); (c) parity order (lower fertility in primiparous); (d) health status; and (e) reproductive rhythm (lower in intensive than in semi-intensive rhythm, thus, lower with inseminations during the first week postpartum than at 11 days postpartum) [11].

Prolificacy: the indexes are conditioned by the genetic origin of the crossbred females. Maternal lines under genetic selection, for example, line Prat, selected for litter size at weaning at IRTA, have nowadays 2.68 weaned rabbits per litter more than in 1992 (personal communication).

Mortality at birth (also known as mortinatality) is usually associated to prenatal survival of the genetic line and mistakes in management, for example, as no nest disposal or stress at first parity (water stress, noisiness) that may increase cannibalism.

General mortality indexes: they are obviously conditioned by the health of the herd, which should be improved by the increase of biosecurity. Some techniques that have been seen to reduce sanitary problems are:

- All-in/all-out system: this technique avoids cohabitation of animals from different ages and physiological states in the same location, and it seems to favor the technical indexes and the reduction of use of antibiotics [9, 12].

- Single batch or multiple independent batch management (the female does not change to another batch when a negative palpation is detected) reduces the incidence of health problems because it avoids rearing of animals of different ages and physiological status in the same location. Moreover, management with one single batch increases the possibility of practicing the all-in/all-out system.

Mortality during the fattening period is mainly associated with digestible disorders, especially since the beginning of the twenty-first century when the epizootic rabbit enteropathy (ERE) appears as a pandemic disease. The ERE affects especially after weaning, when the young rabbit is changing from milk to solid feed with a still immature digestive system and under the stress of the weaning. In these circumstances, the ERE increases mortality and morbidity of the young rabbits. The agent causing the ERE is not yet determined, and the incidence of damage may be reduced with different techniques as (a) variation of feed composition (additives, the reduction of protein level in the feed, etc.); (b) feed restriction during the 2–3 weeks after weaning; and (c) delay of the age at weaning.

Mean slaughter weight: the index can be rarely changed, as it is fixed by the demands of the market. However, the number of days and the quantity of feed needed to achieve that weight can be reduced by using rabbit sired with bucks from paternal lines selected by growth or feed efficiency with lower conversion index during the fattening period. For instance, line Caldes, selected for growth rate in IRTA during the fattening period, presents at 60 days of age a food conversion rate 8.3% lower than a line selected for reproductive criteria [13].

Global feed conversion rate: it is one of the most important technical indexes. It represents the number of kilograms of feed consumed to obtain 1 kg of fryer rabbit for sale and summarizes success in fertility, prolificacy, and survival. The global feed conversion rate should be as low as possible since feed costs represent more than 70% of variable costs. There are many factors that affect this index:

- Health status: morbidity and mortality in all the steps will increase the global feed conversion rate, especially when mortality occurs at the end of the fattening period.

- Mortinatality.

- Food composition:

    - If the prevalence of health problems in fattening is high, feeds with low protein and energy levels prevent digestive problems and could reduce the global feed conversion rate. This involves lower mortality and lower cost per kg of feed, although consumption would increase.

    - If the prevalence of digestive problems is low, feeds with high levels of energy and protein, with a higher price, require less consumption and shorter periods of fattening, which would also reduce the global feed conversion rate.

- Genetic lines: lines selected for growth rate and/or feed efficiency and prolificacy will reduce the global feed conversion rate.

- Environmental conditions: a balance must be struck between savings in environmental control (heating, ventilation, etc.) and feed consumption. A low temperature in the fattening rooms increases the consumption of feed for thermoregulation and therefore increases the global feed conversion rate.

- Semi-intensive reproductive cycles, shorter than the extensive ones, and high fertility rates reduce global feed conversion rate.

## 5. Conclusions

Management is necessary as a source of information to detect weak points for improvement, make decisions, and evaluate the technical and, above all, economic consequences.

There is information on technical management programs (especially in Spain and France), but economic analyses of rabbit production are very rare. Feed is the main cost, and the global feed conversion rate gives an idea of production efficiency, as do other indexes such as the production expressed by insemination that also includes reproductive efficacy. Within our reach, the most important traits related to economic profitability are conversion rate, litter size, fertility, and kitten survival. We must not try to save on genetics, energy, or preventive veterinary measures since they only represent a small percentage of production costs.

## Acknowledgements

Authors are grateful to the Instituto Navarro de Tecnologías e Infraestructuras Agroalimentarias, SA (INTIA), Conejos Gallegos Cogal S Coop (COGAL), and Institut Technique de l'Aviculture (ITAVI) for their excellent and tenacious work in technical and economic management in rabbit farms. Authors are also thankful to Luc Maertens for its contribution in the characterization of the rabbit Belgian market.

## Conflict of interest

The authors declare no conflict of interest.

## Author details

Mariam Pascual[1*] and Ernesto A. Gómez[2]

1 Institute of Agrifood Research and Technology (IRTA), Barcelona, Spain

2 Valencian Institute of Agrarian Research (IVIA), Castellón, Spain

*Address all correspondence to: mariam.pascual@irta.es

## References

[1] FAOSTAT Online Database. 2018. Available from: http://faostat.fao.org/ [Accessed: 20 May 2020]

[2] Cartuche L, Pascual M, Gómez EA, Blasco A. Economic weights in rabbit meat production. World Rabbit Science. 2014;**22**:165-177. DOI: 10.4995/wrs.2014.174

[3] Garcia ML. Embryo manipulation techniques in the rabbit. In: Payan-Carreira R, editor. New Insights into Theriogenology. IntechOpen; 2018. pp. 113-133. DOI: 10.5772/intechopen.81089

[4] Gullón J, Prieto C, Sánchez M, García C. Gestión técnica 2017–2018. COGAL. 2019;**February**:38-40. Available from: http://cogal.net/pdf/2019.pdf

[5] Coutelet G. Résultats technico-économiques des éleveurs de lapins de chair en France en 2014. In: Proceedings of 16èmes Journées de la Recherche Cunicole. 24-25 November 2015; Le Mans; France. 2015. pp. 193-196

[6] Azard A, Lebas F. Productivité des élevages cunicoles professionnels en 2005. Résultats de RENALAP et RENACEB. Cuniculture Magazine. 2006;**33**:92-99

[7] Pascual M, Gómez EA. Base de datos del sector cunícola español. Boletín de Cunicultura. 2008;**156**:58-59

[8] INTIA. Gestión de costes y márgenes 2018: Producción porcina y Cunicultura. Boletín Informativo de Ganadería. **2019**:3

[9] Lebas F. Performances moyennes des élevages cunicoles en France pour l'année 2017. Résultats RENACEB. Cuniculture Magazine. 2018;**45**:22-27

[10] Flemish Ministry of Agriculture, Department of Animal and Fishery. Technical Data and Production Costs of Rabbit Meat Production in Belgium. 2018, internal note

[11] Castellini C, Dal Bosco A, Arias-Alvarez M, Lorenzo PL, Cardinali R, Rebollar PG. The main factors affecting the reproductive performance of rabbit does: A review. Animal Reproduction Science. 2010;**122**:174-182. DOI: 10.1016/j.anireprosci.2010.10.003

[12] Rafel O, Gómez EA, Casas J, Garriga R, Pascual M. El vacío llena: Impacto técnico y económico de los sistemas todo dentro todo fuera (TDTF) en cunicultura. In: Proceedings of the XLI Symposium of Cuniculture. 12-13 May 2017; Hondarribia; Spain. 2017. pp. 112-118

[13] Pascual M, Piles M, Pascual J, Ródenas L, Velasco M, Herra W, et al. Análisis de la eficiencia alimentaria durante el crecimiento de líneas de conejo seleccionadas por diferentes objetivos. In: Proceedings of the XLIV Symposium of Cuniculture. 5-6 Juny 2019; Aranda del Duero; Spain. 2019. pp. 99-103